Praise fc

*Earth Re*

The real and imagined consequences of contaminated soil and water have been some of the greatest impediments to restoration of urban wastelands for food production. Leila Darwish combines the experience of the pioneer activists and innovators with the expertise and knowledge of remediation professionals to create an empowering guide for the large numbers of citizens looking for guidance on this issue.

*Earth Repair* includes enough technical detail and explanation to get most readers up to speed on the subject. The case studies provide empowering examples of how low cost remediation techniques that reflect permaculture design principles can be used to enhance community resilience and advance social justice.

In the energy descent future, many more people will be growing food on contaminated land; out of necessity! *Earth Repair* offers the hope that this can be done without fear of further eroding health and well being.

—David Holmgren, co-originator of permaculture and author,
*Permaculture: Principles and Pathways Beyond Sustainability*

*Earth Repair*, what a brilliant and useful book! Leila Darwish & New Society have brought forth a book for people who will not wait around to heal the world. With a broad and deep view of the historical dynamics of thoughtless upheaval and waste, *Earth Repair* provides thorough, local-action strategies that communities, with or without resources, can undertake to remediate their damaged landscapes. In accessible language, this book explains how to deal with a serious local issue while also shining light on where to go to deal with the source!

—Mark Lakeman, cofounder, The City Repair Project,
communitecture, and the Planet Repair Institute

*Earth Repair* is a fantastic introduction to grassroots bioremediation — an indispensible guide for citizen scientists, permaculturists, and ecological justice activists wanting to proactively address the legacy of environmental pollution which we've inherited from our industrial civilization. Within is a highly accessible toolkit of techniques and skills usable by the average person, empowering them to safely destroy or immobilize common contaminants by partnering with familiar biological allies such as microbes, worms, fungi, and plants. As we transition into a sustainable society, this book will be a key text, critical for informing communities in the process of de-toxifying our planet.

—Scott Kellogg, educational director of the
Radix Ecological Sustainability Center and author,
*Toolbox for Sustainable City Living – A Do-It-Ourselves Guide.*

We are an odd, almost unique, creature that soils its own nest. As we've become more industrially and technologically muscular, our soiling has penetrated into the heart of Earth's systems, where we now pile our filth upon genetics, delicate geochemical balances, and climate. We have destabilized nature, and we won't find a way out the same way we came in. Darwish's good news is that nature WANTS to heal, and even knows how. We just have to use the tools she gives us.

—Albert Bates, author, *The Post-Petroleum Survival Guide* and
*Biochar: Carbon Farming and Climate Change*

# earth. repair

## A GRASSROOTS GUIDE TO
## HEALING TOXIC and DAMAGED LANDSCAPES

Leila Darwish

new society
PUBLISHERS

Cover design by Diane McIntosh.
Image © iStock (contour99)

Printed in Canada. First printing April 2013.

New Society Publishers acknowledges the support of the Government of Canada through the Book Publishing Industry Development Program (BPIDP) for our publishing activities.

Paperback ISBN: 978-0-86571-729-9
eISBN: 978-1-55092-529-6

Inquiries regarding requests to reprint all or part of *Earth Repair* should be addressed to New Society Publishers at the address below.

To order directly from the publishers, please call toll-free (North America) 1-800-567-6772, or order online at www.newsociety.com

Any other inquiries can be directed by mail to:

New Society Publishers
P.O. Box 189, Gabriola Island, BC V0R 1X0, Canada
(250) 247-9737

New Society Publishers' mission is to publish books that contribute in fundamental ways to building an ecologically sustainable and just society, and to do so with the least possible impact on the environment, in a manner that models this vision. We are committed to doing this not just through education, but through action. The interior pages of our bound books are printed on Forest Stewardship Council®-registered acid-free paper that is **100% post-consumer recycled** (100% old growth forest-free), processed chlorine free, and printed with vegetable-based, low-VOC inks, with covers produced using FSC®-registered stock. New Society also works to reduce its carbon footprint, and purchases carbon offsets based on an annual audit to ensure a carbon neutral footprint. For further information, or to browse our full list of books and purchase securely, visit our website at: www.newsociety.com

Library and Archives Canada Cataloguing in Publication

Darwish, Leila

Earth repair : a grassroots guide to healing toxic and damaged landscapes / Leila Darwish.

Includes bibliographical references and index.
ISBN 978-0-86571-729-9

1. Bioremediation. I. Title.

TD192.5.D37 2013          628.5          C2013-901886-7

# Contents

# Many Thanks

To my mother, my father, my sister and my friends: Thank you for your incredibly generous support, powerful inspiration and love.

To the visionaries, activists, healers and earth workers who took precious time out of their busy lives to share their knowledge with me for this book: The gift of your expertise, stories, inspiration and cautionary tales form the basis *Earth Repair*. Thank you Starhawk, Ja Schindler, Peter McCoy, Scott Kellogg, Paul Stamets, Kaja Kühl, Heather Hendrie, Marika Smith, David Holmgren, Anita Burke, Mia Rose Maltz, Robert Rawson, Scott Koch, Freeda Burnstad, Nance Klehm, Jodi Peters, Ferdinand Von Druska, Andrew Butcher, Mark Lakeman, Oliver Kelhammer, Matt Feinstein, Asa Needle, Adam Huggins, Carol Bilson, Jay Rosenberg, Scott Kloos, Elaine Ingham, Guido Mase, Leah Wolfe, Kate O'Brien, and Paul Horsman.

To this living Earth and its wild ones: Thank you for the nourishment, inspiration and sanctuary you have offered me and all those I love. Your resilience and generosity never cease to amaze me.

# Introduction: Manualfesto

*Let the beauty we love be what we do. There are hundreds of ways to kneel and kiss the ground.*

— Rumi

FROM THE CONTAMINATED SOIL IN BACKYARDS, community gardens and vacant urban lots to brownfields, big oil spills and nuclear accidents, how can we transform toxic and damaged landscapes into thriving, nourishing and fertile places once more? How can we respond to environmental disasters in radical, accessible and community-empowering ways?

*Earth Repair* is a grassroots tool kit. It explores some of the visionary and practical tools for healing and regenerating damaged ecosystems — and the powerful and fertile work of reclaiming resilience and hope in our communities. It is an introduction to the dynamic and evolving field of earth repair and grassroots bioremediation. The ideas explored in this book seek to empower and support the work many of us do everyday to grow healthy food and medicine amidst the polluted and damaged soils of our backyard gardens, community commons and wild lands. *Earth Repair* also seeks to support the efforts of emergency first responders, community members and grassroots bioremediators on the front line of oil spills and other industrial accidents whose toxic legacies lay siege to our health and the health of the lands that sustain us.

There is so much potential for this work, if done right, to jump the gap from novel ideas into life-transforming community practices. To move past the narrow conversation on sustainability into the powerful territories of regeneration and resurgence! To speak to people's fears and transform despair into hope, apathy into action. To ensure that no community is reduced to drinking bottled water because their rivers and groundwater have been poisoned, and that no living being has to suffer tragic health impacts because their air, soil and food have been contaminated.

This is the spirit behind this book on earth repair. It permeates everything from the simplicity of building compost or planting sunflowers to bind and breakdown contamination, to the more difficult work of responding to environmental catastrophe and engaging in the fierce resistance work it takes to change the rules of the game so that such catastrophes never happen in the first place.

Many of us have yet to see the work of grassroots bioremediation done in our communities. Maybe we've heard stories or been to a workshop where we've learned that mushrooms can clean up oil spills and plants can suck up heavy metals. But when faced with wounded and contaminated spaces or environmental disasters, we often have no idea how to translate the myths and legends of regenerative solutions into living realities. The stories, ideas, recipes and tools shared in *Earth Repair* are powerful offerings from the many community responders, permaculture healers, grassroots bioremediators and environmental restoration workers who are attending to post-disaster landscapes. By reconnecting with living beings, living cycles and the living earth, they are bringing life back to city lots, backyards and our damaged wild spaces and hearts.

The information and stories presented in *Earth Repair* are just a small taste of the vast field of fertile knowledge waiting to be explored and engaged. May it inspire you to initiate and support all those great projects, big and small, that will repair, reconnect, revitalize, regenerate and re-wild the many broken and poisoned landscapes in which we live.

# CHAPTER 1

# Roots of Repair: Decolonization and Environmental Justice

BEFORE LAUNCHING INTO THE WILD WORLD OF EARTH REPAIR and grassroots bioremediation, I would like to take a moment to address the intersection of decolonization, environmental justice and earth repair. As many of us are seeking to heal the lands we call home, we must recognize that these lands were stolen from Indigenous peoples and devastated through acts of colonization. If you are not indigenous to these lands, then you are participating in and benefiting from an ongoing crime against the people of these original nations. Though we cannot choose where we are positioned at birth, how we relate to that positioning is absolutely critical in addressing and challenging historical and current injustices. We have a lot of work to do in taking responsibility for these injustices and ensuring they do not continue. We must find honorable and meaningful ways to move forward that repair our relationships with each other and the land. As the author of *Earth Repair*, I feel that it is critical to acknowledge this reality and to ask that you keep it in mind as you carry the information within this book out onto the lands beneath your feet.

These lands not only bear the wounds of an industrial war waged upon their forests, rivers, meadows and mountains, but also the deep painful history and ongoing practice of colonization with its continued disrespect, violence, occupation and environmental destruction forced upon Indigenous peoples. The Indigenous peoples who traditionally tended these lands and the many who continue to do so never willingly gave up their lands. They have been forced from them so that ports, forts, malls, housing developments, roads, farms, public gardens, golf courses, car dealerships, schools, office towers, mines, dams and pipelines could erase and bury the vibrant life-support systems and strong communities that existed here previously.

One of the ways that colonialism seeps into the everyday work of earth repair is through the dark legacy and enduring tradition of environmental racism. Indigenous communities, people of color and low income neighborhoods are all too often sites targeted for heavy industry, military bases, waste dumps and higher levels of pollution. People in these communities suffer more health and environmental impacts than their affluent, predominantly white neighbors. Whatever laws, agreements and regulations that may be present to challenge such injustices are often ignored and violated by corporations and governments. When it comes to recovering from environmental disasters and industrial accidents, these communities often receive little notification, support or effective cleanup, if any at all. This reality also applies to poorer countries, which have been traditionally used as dumping grounds, and whose rich environments have been exploited and destroyed by wealthier countries and their corporations.

What does this look like on the ground? Massive tar sands mines that are poisoning the world's third largest watershed and the traditional territories and communities of the Cree and Dene peoples. Health impacts suffered by the Aamjiwnaang First Nation community which is surrounded by the many refineries of Chemical Valley in Sarnia, home to Canada's largest cluster of chemical, allied manufacturing and research and development facilities. Low income and/or racialized urban neighborhoods whose homes, schools and gardens sit atop contaminated soils or in the midst of heavy industry. The list goes on. Though some folks like myself can simply relocate to cleaner and healthier living spaces, many people cannot, due to economic circumstances as well as community, cultural and historical connections and responsibilities that hold them to these very places. In these realities, environmental destruction, tragic health impacts, oppression and social justice all collide, which is why some folks do the work of earth repair for the love of the planet, some for the love of their loved ones and community and many for both.

The practice of earth repair not only involves detoxifying and revitalizing the land by working with plants, mushrooms and microorganisms; it must also include the powerful work of decolonization that seeks to deeply repair and enliven both the ecosystems and the communities that support thriving natural systems.

So as you consider initiating and engaging in earth repair and grassroots bioremediation work, acknowledge, respect and honor the original stewards of these lands and their indigenous rights, life practices, knowledge, resistance and sovereignty. Ask permission for the earth work you are about to do.

Find out who your ancestors are, where your people come from, the ways they cared for and honored their ancestral lands and at what point this relationship was lost or destroyed. Try to see the land around you with new eyes, the kind that can see beyond the grid systems, megaprojects, dams, fields of monocultures and urban jungles. Take the initiative and responsibility for educating yourself on the buried and hidden stories of the lands you call home and the peoples you share them with. Acknowledge and actively work to challenge the power inbalances and destructive ideologies that have created situations whereby the ability for many different peoples and communities to be healthy, have right livelihoods and live in positive relationship to Earth have been severely compromised. Challenge the capitalist culture that sees the Earth and its living beings as something to be owned, commodified and destroyed for profit, and do not replicate it in the earth work that you do.

Finally, ask yourself how you will meaningfully undertake the deep and life-altering work of decolonization and the powerful solidarity and resistance work which it requires. We must challenge the voracious colonial processes that created and continue to create such damage in the first place. Because without that, your healing work for this planet will only be superficial and shortlived at best.

CHAPTER 2

# Earth Repair and Grassroots Bioremediation

MOST OF *EARTH REPAIR* will focus on grassroots bioremediation, but it's important to remember that bioremediation is part of a much larger practice of earth repair, earth care and right relation. There are many other tools and practices that fall under the broad and deep scope of earth repair. In this book, we are just scratching the surface, focusing on a few tools for cleaning up contamination, tools that also renew and regenerate damaged landscapes.

Bioremediation works with living systems to detoxify contaminated environments. It includes microbial remediation (engaging the healing power of microorganisms like bacteria and fungi), phytoremediation (engaging the healing power of plants) and mycoremediation (engaging the healing power of mushrooms) to heal contaminated and damaged lands and waters.

To me, true grassroots bioremediation and earth repair aspire to the following:

- community accessibility and affordability
- working with nature to assist in its healing
- DIY (do it yourself) remediation, restoration and regeneration techniques that are high impact, low input, non-toxic, simple and easy to replicate
- prioritizing deep ecological healing and community justice as the motivating force

- embedding the skills and necessary infrastructure in our communities, especially communities most at risk for environmental disasters and contamination issues
- applying whole systems, multi-kingdom, multi-species, multi-tool approaches
- honoring local knowledge, local resources and engaging local decision-making
- empowering people to directly respond to their circumstances and crises in ways that increase their knowledge and self-determination and result in real improvements for the planet and their communities
- avoiding externalizing the problem whenever possible
- challenging and actively resisting the privatization and corporate ownership of solutions, spaces and living beings
- engaging in the powerful preventative medicine of resistance through grassroots organizing and mobilizing to stop destructive projects from going ahead
- acknowledging that it takes a community to make an earth repair project successful and seeking to build and maintain respectful relationships and trust with our many living earth repair allies (bacteria, plants, fungi, animals, people, ecosystems).

Grassroots bioremediation and earth repair mean choosing our interventions wisely, observing nature, learning and drawing inspiration from its brilliant design. They involve figuring out how to restore cycles and natural processes that have been interrupted, and working with nature's own healing mechanisms to organically and holistically restore wellness to a site. This work is not about being dogmatically attached to a specific tool or method. The true grassroots bioremediator is not someone who believes that mushrooms will save the world and wants to apply them to every situation. Instead such a healer is willing to consider that we have many tools at our disposal, that a lot of these tools work best in concert or in succession and that the environmental conditions of the sites we are working with may favor one tool over another. There is no one-size-fits-all approach.

This way of doing earth repair does not make much profit for

# Conventional Remediation

Conventional remediation refers to the many ways that industry and government are approaching the removal of harmful chemicals and metals from contaminated sites, specifically looking at how they repair or "clean up" contaminated soil and water. Though I have some criticisms of how conventional remediation is done, it is important to recognize that there are situations where it can achieve strong results, some that may not be achievable by grassroots bioremediation, due to the extent and extreme nature of the destruction caused by the industrial system.

Both conventional and grassroots remediation have their blind spots. Some grassroots bioremediators can be blissfully overconfident about the healing power of nature without understanding the nature of contamination on a site. Some conventional remediation professionals can possess an overconfidence in using technology, chemistry and engineering-based solutions that do not work with the innate ecological intelligence of the land or the people who reside most closely to it; they can fail horribly at restoring the ecological integrity of a site. With the almighty dollar, public relations and government regulations as primary drivers in most cases, solutions and treatments are often selected based on their cheapness and expediency, leading to more of an excavate, cover and bury approach to cleanup.

That said, it is important to build relationships with key allies in this line of work. Many folks working in conventional remediation are in it to do good work and are equally unsatisfied with the outcomes and compromises they are forced to make by the corporations who call the shots. They have a lot of experience working with extensively damaged land and big projects, and this important knowledge and skills could be helpful to increasing the effectiveness of grassroots bioremediation efforts.

In Appendix 2, you will find descriptions of different conventional remediation technologies. I highly recommend you familiarize yourself with them. As a grassroots bioremediator or as anyone living in a community dealing with contamination, it is important to know what technologies and tools are currently being applied to cleaning up contaminated sites and what their impacts truly are.

industry. Its implementation pays off over time. Earth repair also doesn't cover up and hide the problem quickly enough to save public relations face or to get a speedy stamp of approval so that the problem is all wrapped up and companies can move on. Unlike most conventional remediation that treats the symptoms, patents the "cures" and profits from the lack of ecological health, earth repair methods seek the roots of the illness and act in ways that restore overall health and balance to these natural systems. These methods require frequent and lighter interventions over a longer period of time. These tools for repair and regeneration are not tools fashioned from the same industrial paradigm that created the problem in the first place. Unlike the chemicals and heavy machinery used in conventional remediation, you are working with living beings to restore life and health to other living beings. This is complex work that can be hard to control and hard to command with the quickness often required for disaster response.

These different microbes, plants and mushrooms grow differently in different soils, microclimates and in the presence of different contaminants. Therefore applications need to be tailored to specific site characteristics and conditions, making it challenging to have ready-to-go stock on hand. Some of these living beings are more sensitive than others, and you will have to coax and care for them in order to get them to the point where their healing powers are unleashed. In a way, you may have to become bacteria, plant and mushroom whisperers, and those of us who combine that care with a skillful and detail-oriented approach will be the most successful at this work.

## Grassroots Bioremediation Reality Check

When I set out to write *Earth Repair*, I wanted to compile a how-to guide of empowering, accessible and holistic tools to help clean up the mess that's been made of this beautiful place we call home. The reality may be that we just aren't there yet. These tools are still being developed and honed; many of them are facing resistance and inertia from governments, industry and some professionals, which is further delaying their growth and ability to succeed. In order to be effective at this work, it is important to know what are some of the barriers you may face in doing grassroots bioremediation.

## Barriers to Community-Based Bioremediation

BY SCOTT KELLOGG

The urban gardener is in regular physical contact with soil, breathing its dust and eating foods grown from it. Few others have such an intimate relationship with city soils, a resource that is seen by most others as something that only serves as a foundation for buildings and roads. Logically then, gardeners are concerned with soil contamination issues and are looking for simple and low-cost means to address them.

For a number of years now, urban gardeners and their supporting organizations have been aware of the concept of bioremediation. Bioremediation's use of naturally occurring organisms, apparent affordability and minimal disturbance to soils all add to its attractiveness. The idea of partnering with life-forms such as bacteria, plants, worms and fungi (all of whom gardeners are already familiar with) greatly adds to its appeal. Numerous scientific studies have been conducted that support bioremediation's effectiveness — there's no question that given the right circumstances, these organisms have the potential to degrade, immobilize or sequester a variety of contaminants. Bioremediation would appear to be an ideal and elegant solution to issues of soil toxicity. Why then, have bioremediation techniques not yet been put into use broadly as a means to remediate contaminated soils in urban gardens? Why are we not seeing citizen groups applying the tools of bioremediation and publishing their results?

These are questions that I have been asking for a number of years in my work designing ecologically and socially regenerative systems in urban environments. In 2004, the Rhizome Collective, an organization that I co-founded in Austin, Texas, received a $200,000 grant from the US Environmental Protection Agency to clean up a brownfield site located in the city. This cleanup primarily involved removal of tons of trash and debris — a twisted mountain of concrete, rebar, wood scraps, tires and carpet scraps — from a former illegal dumping site. Although levels of organic and heavy metal pollutants on the site met residential standards, concerns remained about the safety of gardening there, post-cleanup.

Unfortunately, funding in the grant would not cover the cost of soil remediation. Shortly afterwards, I was part of an effort to establish a community-based bioremediation plan in post-Katrina New Orleans to help address residual hydrocarbon contamination left behind from the storm. Compost teas were applied to areas known to be affected by pollutants. While the program received donations of services from soil testing labs, the funds were insufficient to carry out a properly managed remediation program on the scale that was necessary.

The barriers to community-based bioremediation are many. One such obstacle is that there is still a great deal of mystique, particularly to the less scientifically literate, surrounding bioremediation and its processes. This lack of understanding can make bioremediation an intimidating prospect to many. The vast majority of literature concerning bioremediation exists in scientific journals, written in a dry academic style that is close to unreadable by the layperson. The bulk of these studies are conducted in highly controlled, sterile laboratory conditions, incredibly different from the diverse, heterogeneous and competitive ecologies that exist in a garden environment.

In order for this boundary to be spanned, a few individuals that are scientifically literate will need to wade through the journals and distill a series of guiding principles and best practices usable by the average gardener. From there, a push needs to made from within and outside of academic institutions to conduct a greater number of field-based trials, where proven bioremediation techniques are put to the test in real-world conditions. Emphasis needs to be placed on techniques that are simple, affordable and that make use of commonly available biological agents. The focus of these studies should be on the top 12 inches of soil, the zone in which the majority of urban gardeners are active. Partnerships between citizen groups and academic institutions are vital, as universities have access to technological resources that gardeners are commonly lacking but are necessary to conduct such trials.

Cost is another significant barrier to implementing bioremediation techniques. While bioremediation is relatively inexpensive, especially when compared to more intensive means of conventional remediation, its use still requires some expenses, particularly when soil testing is involved.

For example, spreading spent mushroom substrate over an oil spill could conceivably be done for little or no cost. Testing the contaminated soil, however, to be sure that the total petroleum hydrocarbons have been reduced to safe levels can be prohibitively expensive — potentially hundreds or thousands of dollars. While the spent substrate application may have been successful in degrading the oil, without verifiable data to prove its effectiveness it is difficult to get the support needed to replicate the process on a broad scale.

Governmental agencies ideally would play a role in funding citizen-based cleanups, although they have done very little to date. Most of the government funds that exist for brownfield remediation go to large-scale developers, who are primarily interested in meeting regulatory obligations as quickly as possible. These developers typically favor a "dig and dump" approach to soil remediation, rather than dealing with the longer time scale and other uncertainties that can accompany bioremediation. Additionally, many governmental agencies charged with environmental protection are reluctant to work with citizen groups, fearing liability were something to go wrong. Consequently, governmental agencies are commonly unfamiliar with small-scale bioremediation. It is my belief that it is very much in the interest of government agencies to alter this policy. As interest in community gardening increases, so will the number of people wanting to partake in soil remediation. These people are going to attempt remediation, whether or not they receive assistance from agencies. Therefore, it makes sense for agencies to offer some form of guidance or assistance so that people do not put themselves in harm's way.

Part of this work would be developing a solid method of risk analysis for exposure to soil toxins. Currently, no such framework exists. It is important to be able to answer important questions like "what is the danger of being exposed to a particular contaminant in the soil, and further, what is the danger of that being taken up into a plant and being passed into my body?" These aspects of risk need to be weighed against all the benefits of gardening, such as improved nutrition, physical exercise and enhanced community relationships. Developing such a framework is a multi-disciplinary task, involving the fields of ecology, toxicology, public health and medicine.

Developing a protocol for qualitative soil analysis is another innovation that could reduce the cost of bioremediation. It may be possible to create a method for assessing the quality of soils using what are called *bioassays*. An example of a bioassay would be testing seed germination rates in soils with known levels of toxicity. Using this information, it could be possible to determine contaminant levels in soils using only plant seeds, potentially cutting the cost of soil testing dramatically.

Bioremediation holds great promise for urban gardeners as a tool for achieving improved soil health. Hopefully, in time and with the cooperation of institutional entities, it can go from being an experimental technique to a broadly utilized strategy.

> *Scott Kellogg* is the educational director of the Radix Ecological Sustainability Center in Albany, New York (radixcenter.org) and author of the book Toolbox for Sustainable City Living.[1] He has recently completed a Masters degree in Environmental Science from Johns Hopkins University, where he wrote his thesis on the topic of low-intensity, community-based bioremediation techniques.

Before we dive in to the how-to section of the book, there are a few disclaimers and words of caution to be shared.

## The Importance of Safety

Your personal safety and that of the people you are working with is incredibly important. Depending on what sort of earth repair and grassroots bioremediation work you are doing, you could be putting yourself in a dangerous situation with potential health risks that could impact your life and the lives of those in the community around you.

A lot of government agencies and professionals discourage community folks from doing this work; they may be genuinely concerned for your well-being and do not believe that you are properly prepared and aware of how to handle such a situation. They do not want folks running into dangerous situations ill-prepared, and this is a legitimate concern and one that you should take seriously.

Your health is important and is as important as this work; besides this work is a lot harder to do if you are unwell. Take the proper safety measures. Wear protective gear if you can. There is more information about what that is later in *Earth Repair*. If you do not know what the risk level is or what you can do to protect yourself, do some research or ask a trustworthy professional. Though be suspect when asking the corporations that caused the problem, as they have a history of downplaying risks and not properly protecting community folks.

Just to be clear — this isn't always high-risk work. It depends on the degree and type of contamination of the site or the nature of the spill you are working with. Please err on the side of caution.

### Empowerment and Recklessness

Co-founder of the Rhizome Collective Scott Kellogg warned me about what he called the fine line between empowerment and recklessness when it comes to grassroots bioremediation. Yes it is important to empower folks to get out there to effectively engage in cleaning up, restoring contaminated land and responding to environmental disasters. But we also have to be mindful of getting them all riled up to engage in situations that may be beyond their scope of expertise and present dangerous risks to their health and the community around them. Furthermore, if folks enter into this work on false promises of quick successes and simple solutions, when they face the more complex and messy realities on the ground they will be more likely to give up instead of rolling up their sleeves and shouldering forward.

The truth of grassroots remediation is that it is a far cry from the instant promises of flashy green solutions. Nor is it properly represented in the serene stories of sunflower fields, bubbly brews of compost teas and oil-eating mushrooms. Mushrooms eat oil and help us repair damaged landscapes. It is absolutely correct that microorganisms and some plants can break down chemicals and bind heavy metals. They have always done this powerful work for the planet; it is an essential role they play in the symphony of life and the chaos of catastrophe. But it's just not as quick or simple as some folks make it out to be.

We are working with landscapes and sites that have suffered for a long time, and the pollutants you are working to clean up and the damage you are trying to heal have built up over decades and will not vanish overnight. There is no solution in the remediator's toolbox that can offer that quick fix. Though conventional remediation may offer a few, they are only temporary fixes that clean up one area by shuffling around the contamination to another sacrifice zone, landfill or incinerator. There is no perfect or pure solution, and there are many compromises that we will have to make in this work. That is just the way it is given the nature of the damage we may be dealing with.

We are also working within a system that often does not respect or encourage the work of grassroots bioremediators (though there are many allies and some places where the system and mainstream power players are experimenting and engaging with this work). You will likely be discouraged from doing it, or at the very least have to jump through many hoops to prove yourself, your project and your plans. Embrace this and use it as an opportunity to make your work better and to clear the trail for those who will follow you.

Be bold but don't be arrogant and underestimate the work involved. Invoke temperance and humility, and ask for help when you don't know what you are doing. Be willing to accept your mistakes and adapt your approach and systems. Create a culture of openness so that we can figure out what is working and what isn't, enabling us to learn from each other's successes and mistakes. Start small and make incremental changes, taking time to observe and allow for room to adapt and shift. If you work in this way, you will have less of a mess to clean up if things go awry.

### My Expertise

I am not a remediation professional. I have never worked for any remediation company, been a clean-up worker on an oil spill or remediated a piece of contaminated land. I am on the same learning journey as many of you are and am doing my best to figure out the path. I do so from a place of passion, and with as much awareness for the necessary precautions as I can, given my lack of professional exposure or formal training in this field. My background is in community organizing and environmental issues, permaculture, organic agriculture and herbal medicine. I have an undergraduate degree in Environmental Sciences, which exposed me to some soil science, hydrology and land reclamation, but definitely not enough to be an expert.

A lot of the information in *Earth Repair* therefore comes from interviews I have done with folks who are doing this work in the field, as well as papers and articles I have found. So the information provided is from the experiences of others; I can't promise it will always hold up to be true or effective. As we figure out new and better ways to do things on the ground by attempting the many different forms of earth repair, and as experts open up to sharing more of what they know with the grassroots, I hope that the information provided in this book will be quickly improved by the contributions of others.

### Skilling Up to Scale Up

There is so much that we need to learn in order to do this work effectively. We will not become experts overnight, especially if we are too intimidated and overwhelmed to even take that first step. *Earth Repair* is a series of first steps. It's the low-hanging fruits of grassroots bioremediation. The stuff you can do without making too much of a mess or taking on too much risk, that can serve as tools when all other tools are removed from your grasp. And hopefully the thoughts, ideas, tools and stories captured within its pages will inspire you to learn more so that you can become truly effective in whatever piece of this heavy load you decide to take on. Look to the landscape around you and the community that holds you, and then determine what is needed at this time and how you can fill that need as quickly and earnestly as possible. What knowledge and what skills do we need to learn in order to do this work meaningfully and effectively? What relationships need to be built, what mentorships need to be offered or sought out?

It might mean that some of us have to go back and relearn chemistry or soil science, or strike up friendships with those folks who know how to do soil testing and who do professional bioremediation work. It's time for us to step up to that plate. To really consider how we are going to arm ourselves with the information necessary to truly understand what we are up against so we can be as effective as possible in the healing work we do.

### Patents, Trade Secrets and the Privatization of Solutions

When I set out to write *Earth Repair*, I was completely unaware that bioremediation solutions and species are being patented. This unsavory reality poses challenges to the grassroots bioremediator. From what I can gather, some academics, scientists, universities and remediation companies have patented certain bioremediation solutions, meaning that they believe (and so does the legal system) they "own" these solutions and the rights to use them.

In the true spirit of disaster capitalism, a lot of companies and individuals are trying to make money off destruction, and some corporations can be quite aggressive with scooping up and patenting new clean-up technology. When it comes to using different bioremediation tools that may be patented, professional bioremediator Robert Rawson had the following advice: "An individual who is not commercializing something can use any patent out there. You can go online and find the patent library and read up on different patents and see how they did it and try it yourself. There is nothing illegal about that. What is illegal is to take that and start making money off it."[2]

Some folks believe that patents mean that people who want to use them need to either pay patent holders or ask permission, otherwise they could likely sue you. I was told that this isn't necessarily the case, but patents definitely warrant further investigation. Much may depend on the type of patent involved or the possessive nature of the patent holder.

So be leery of whom you do business or share your research with. It's important that we all fight hard to keep bioremediation solutions and living beings in the commons where they belong. At the end of the day, it is wrong for any individual or corporation to say they have purchased or earned the exclusive rights to another living being or to a lifesaving tool. At the same time, we live in a world where corporations can snap up these tools, patent them and suppress them; so many scientists and sympathetic remediation folks are patenting what they discover in order to retain public access.

I also didn't anticipate how my efforts to make available the secrets, recipes and practices of the folks doing the research and the remediation work on the ground would get blocked by the business of it all. The reality is that remediation is a business, and some big companies are in it for maximum greediness and profit, while other smaller companies and professional earth repair folks are in it to make a decent living. Finding experts willing to sit down with me and let me copy their methods down was challenging. Due to the competitive nature of the field, they don't easily share them because if they give out what makes them good at what they do, they are writing themselves out of a job and their business loses its advantage. I understand that, but I wish it were not the case as I don't think it does any favors for community folks on the ground.

### *Organic vs Inorganic Contaminants (Chemicals/Compounds vs Heavy Metals/Elements)*

It is important to keep in mind the difference in working with chemicals versus heavy metals in bioremediation. You need to consider which tools in your toolbox to use, depending on what sort of contamination you are dealing with. Organic chemicals (such as hydrocarbons and pesticides) can be broken down or transformed while heavy metals cannot. With organic chemicals, you are aiming to break them down or transform them into more benign substances. With heavy metals, you are seeking to either extract them from the soil by isolating and sequestering them so that they can be concentrated in one spot or removed from the site for safe disposal, or to bind them in the soil so that they become less able to be taken up by plants or moved into the water table. You cannot break down or transform a heavy metal; it will always be present, either in a different location or in a less mobile and reactive form. In some cases, you may find yourself dealing with both organic chemicals and heavy metals (for example different fuels and pesticides have heavy metals in them as well as chemicals).

## Who Is Who in the Toxic Zoo?

When it comes to cleaning up contaminated sites, understanding what contaminants are present, how they act and their environmental and health impacts allows us proceed as effectively as possible. Knowing what are the major sources of these contaminants also allows us to cast backwards and determine what may be present on a site depending on past uses and activities.

To fully understand how substances act and react is complicated. Some metals are more reactive and move through the soil more easily when the soil pH is acidic, others when it is more basic. Some chemicals leave the body quickly after a person or animal is exposed. Others (like the persistent organic pollutant, DDT) may remain in fat, blood or bones for a long time. Chemical mixtures may be more harmful than the individual chemicals. In oil spill cleanup, it is often the deadly combination of solvents, dispersants, crude and saltwater that can cause severe health and environmental impacts for the environment, nearby communities, clean-up workers and volunteers. Taking a chemistry class wouldn't be a bad idea if you are serious about dealing with contaminants. Finding an ally who has this knowledge would be equally helpful.

Please refer to Appendix 1 for information on different heavy metals and chemicals that you could find on sites you are working with, as well as information on what industries and activities create them and their health impacts. ☞

It is also important to remember that there are different standards for what con-stitutes acceptable and safe concentrations for each contaminant, and these standards differ from place to place. Some health professionals believe that there are no safe levels for certain contaminants. Due to a pretty strong chemical and industrial lobby, the standards do not always reflect what is best for human and ecological health. When looking at the safe levels of a contaminant, remember that it has been found by some studies that though low levels of certain contami-nants in an acute or short-term exposure will not cause adverse impacts, those same levels with prolonged or chronic exposure might.

### Permission vs Forgiveness

For folks who like to or have to stay on the right side of the law, you'll want to find out who owns the site you want to remediate and check in with your local city environment office to see what hoops you need to jump through and what approvals you need to get. It will likely involve many meetings and presentations, finding your way around liabilities and through bylaws, regulations and other bureaucratic red tape. You may have to get an engineer or remediation professional to approve your remediation plan. As most earth repair projects take several years, developing a good relationship with the landowner, community and local government helps to lay a strong foundation for collaboration, success and security. However, going this route can also tie you up in a lot of red tape and delay your earth repair work substantially.

Many organizers and grassroots remediators who are doing the work of transforming land from brownfields or greyfields to greenfields in communities across this continent have to work with laws and compro-mise their pace and approach as a result. This is the on-the-ground reality that no amount of idealism or fist pumping can change. It is a reality that has allowed their projects to succeed and remain in place over time.

I interviewed Mark Lakeman, visionary architect and co-found-er of the City Repair Project, which is a community initiative best known for their creative community transformations and reclama-tions of public spaces in Portland, Oregon, USA. Every year, City Repair puts on the Village Building Convergence, which is ten days of community work parties and workshops on different sites around

Portland, transforming the fabric of the city block by block through permaculture, natural building and placemaking projects. City Repair first started its work with a mobile community tea house installation and then followed that up by taking over an intersection to reclaim and transform it into a village square. They had no permits for either action, and they were definitely not the favorite child of the city administration in the beginning. But today, you'd never know it.

I asked Mark more about how City Repair approached dealing with bureaucracy and what allowed them to develop a more positive relationship. Lakeman explained, "The trick was to get them to identify with our initiative. Every city and town has goals and objectives, like 'let's make the streets safer, slow traffic, etc.' Dig into the website of your municipality, find their vision statement and then find all the language in there which relates to your initiative. Don't just go to them and say 'we are going to sprinkle mushrooms on the ground here' — they won't understand that. If you start with the form and they are not familiar with it, they will react and tend to say no. So first ground the conversation in common goals. Say to them 'there are these five things that the municipality has identified that it wants to do, and we will accomplish all of those things with this project.' Then describe the method or form of how you want to undertake it. Basically you are making a proposal and outlining the way you are going to do it. As soon as a bureaucrat can hold a piece of paper, it all becomes real for them."[3]

If time is of the essence and you are pretty convinced you will receive only opposition and delays from the government, you may want to get all guerrilla with your earth repair, squirrelling away trash, spraying compost tea and planting mustards and sunflowers in your wake. You may choose to proceed without waiting for the bureaucracy to catch up because you are strong and solid with your group and community and feel confident to deal with the government or the company when it butts up against you. You may feel pretty empowered by what you are doing and may be in a situation where you want to set a precedent for the fact that people have a right to do this work — to revitalize community and Earth. If that's the case and you have no faith or taste for government process or recognition of corporate ownership, go for it.

I'm not advocating that you do anything illegal, but if you, your collective or community decide that you need to move forward in a

certain way because of the experiences you've had or the beliefs you
hold, then that's your prerogative. Just remember that there are legal
risks and health risks. If you want to proceed without permission you
will have to take them into account and also bear in mind that potential conflict could result in the untimely end of your project.

Personally, I envision a time in the not so distant future when governments and companies will have other things to do than to focus
on who is planting what on which abandoned site, and you as a community of folks who have to live near that site or collect water near it
or need more growing space for local food will be able to approach the
situation unencumbered by bureaucracy. We aren't there yet, but one
day we might be.

In deciding how to proceed, we must question whether the bureaucratic processes that we are being asked to follow are there to help or
hinder. Do they create unjust situations that hold the interests of elites
and of industries over the health of the planet and the well-being of the
people who are directly impacted by the decisions made locally? Or are
they legitimately there for public health and safety and because experts
are needed in these cases? In the end, every situation is different, and
the call is yours to make.

## Forward We Go!

Nature has shown us, since the beginning of time, that bacteria, fungi
and plants have played their role with true genius when it comes to
disaster response, recovery, restoration and regeneration of the planet —
just on a different time scale than our modern technological world is
accustomed to. At the end of the day, we might not have all the information and assurances we need, but let's view that as a challenge to
step up to. It's time to get moving! *Earth Repair* is about doing the best
you can as creatively and safely as possible in unfortunate situations
given minimal resources and limited information.

The road is made by walking, and the next few chapters are an in-
vitation to pick up the pace and start trying different methods out. To
start experimenting in your locality, making mistakes, unearthing brilliant new discoveries, finding our own ways forward and being inspired
by the work of others on how to make these earth repair tools ever
more applicable and accessible to our bioregions and our communities.

# CHAPTER 3

# Getting Started

Before you roll up your sleeves and get your hands dirty with different bioremediation tools, there are some things you need to do first to get your site ready.

## 1. Organize an Earth Repair Team

Many hands make for lighter work, and many minds make for brilliant problem solving and designs. Seek out committed and active friends and allies to form your dream team. Keep your group open so that you have enough human power to do the initial and maintenance work, and to avoid burnout! What expertise and knowledge is necessary to the collective in order for it to be productive, dynamic and self-sustaining — so that it meets the needs of its members and the ecologies it is seeking to support? Do you have those skills yet? Who does?

The whole jack-of-all-trades master of none idea won't do you too much good here. We need to develop into a jack-of-all-trades and master of one at the very least. And just as nature abhors a monoculture, seek to create or engage with a team of diverse and dynamic folks with complimentary differences when it comes to skills, so that all niches are filled.

Some of the skills you may want to master or seek in others include:

- phytoremediation
- plant propagation and cultivation
- fundraising and grant writing

- soil science and soil testing
- hydrology
- chemistry and contaminants
- permaculture design
- irrigation
- biology and ecology
- microbial remediation and microbiology
- compost making
- vermicomposting
- compost tea making
- mycoremediation
- mushroom cultivation
- water filtration and treatment (constructed wetlands, greywater systems, floating islands, mycobales)
- public outreach and community engagement/community organizing
- government and/or company relations (proposal writing, lobbying)
- environmental regulations and law
- remediation regulations and standards
- health and safety
- disaster response
- oil spill response
- alternative healing modalities as preventative protection and medicine for chemical and toxic exposure

Consider the time line of the project and what skills will be needed at what times. Be open and ready to reach out, learn or engage those people and skill sets when the time is right. Integrate mentorship, culture building and skill sharing to increase the ecological and social literacy of your community. Let each one teach one. Engaging in a culture of skill sharing and mentorship is key to the inclusivity, growth and accessibility of an earth repair movement and provides a vibrant alternative to more elitist and hierarchical educational and conventional remediation circles.

## 2. Do Your Research about Ownership and Site History

Not all vacant lots are contaminated, and not all contaminated lots are vacant. How will you be sure about your site? If you aren't working in your own backyard or on the land of someone you know who has invited you to be there, take the time to figure out who owns the site; this is key to being able to legally engage the site, as well as to make sure your amazing earth repair efforts won't get cut off by sudden pop-up "landowners" who decide that they want to pave the site and put up condos. Who owns the site is sometimes quite obvious; other times you will have to contact your local government, go through city records to figure it out or ask the elders in your neighborhood. It's up to you as to whether you want to contact and engage landowners or not.

You may be into more guerilla remediation and of the mind that it is better to ask for forgiveness than permission. In some neighborhoods that have been marginalized and neglected by their local government and where there are exceptionally high numbers of vacant lots, you may choose to proceed without going through the standard bureaucratic channels, as both landowner and government are largely absent. Proceeding without permission from the authorities is one way to do it, but if you go that route, you may encounter frustrating obstacles or brick walls later in the game from both the site owner and the government, especially if your site is in an area of interest to them.

Another really important area of research is site history: finding out past uses on-site, what sort of activities have taken place, what sorts of contamination those activities could have left behind. Doing this research will give you a good idea of what contamination problems you may be dealing with and what earth repair tools are best suited to the task.

## 3. Do Soil and/or Water Testing for the Site

In order to know what contaminants you may be dealing with and where they are on-site, you should conduct soil tests and water tests (if your situation warrants them). Soil testing seems to be most commonly needed for urban and rural projects, but depending on the site and future uses, water contamination may be a big issue that needs to be addressed. Different soil tests search for different contaminants, so if you are looking only for heavy metals or organic pollutants, let the soil testing lab know. Also, most soil testing labs return results which are hard for non-scientists to understand, so you may either have to find someone at a local university or government office to explain them to you or ask the lab for explanations. Testing can be quite expensive, depending on whether or not you can get it cheaply or for free from universities or government. If you have to approach a lab yourself, you will end up forking out some serious cash, but this may be necessary. Testing helps you know what you are dealing with and where it is located so you can determine the treatment methods needed.

A great idea to consider implementing in your community is to create an interactive soil map that can be used by different grassroots remediators. Such a map would allow folks to display the results of different soil tests conducted on different sites in your area. By sharing that information, people won't have to waste money testing sites that have already been tested, and we can start building a stronger understanding of the state of the earth beneath our feet. There is often a lot of turnover in community gardening, urban farming and guerilla remediation, so it would be really useful to share site information through some sort of interactive and evolving map or forum. Find more info on soil testing in Chapter 5. Also, refer to Appendix 1 for more information on different contaminants you may encounter on-site.

## 4. Conduct Several Site Assessments

Visit your site several times to get to know it and collect vital information. Observe its patterns, the different forces that flow through it. Try to answer the following questions:

- What fundamental functions have been lost on this site? List the different functions and cycles that take place in the land base/ ecosystem you are working with (be extensive — include social ones) and then observe at what levels they are all present and functioning.
- What work/interventions will bring back and reconnect these cycles?
- Are there signs of bird and animal use — tracks, nests, poop?
- Do people like to cut through or use the site? If so, you will need to take this into account in terms of safety during remediation and maintenance of the site.
- How does water flow through the site? Does it run off? Where to? Does it pool? Where?

- How does the soil look? Are there worms in it? What is the soil texture? What is the soil pH?
- What plants, if any, are growing on the site? Where are these plants? Where are there bare spots?
- Are there debris piles? Where and what is in them? Are they safe to move? Are there any strange substances on-site?
- How does the sun move through this site? Where is there sun, where is there shade and for how long?

These are all critical pieces of information that will allow you to come up with a more successful remediation and regeneration plan.

## 5. Figure Out Your Remediation Treatment Plan

So, you've figured out what sort of contamination you are dealing with and you've taken some time to get to know your site and its conditions. Now you need to consider how you will deal with the contamination and damage, which grassroots remediation tools to use and in what order. Many of the grassroots remediators I talked to recommended a multi-kingdom approach if possible. So basically find a way to work with microbes, plants and fungi. Not necessarily at the same time, but the fact is that they feed each other and feed into each other. Maybe you need to engage one organism to break the big contamination down into smaller pieces, therefore allowing a different organism to take over more efficiently. Sometimes this is the case, other times your site makes the call for you, as it may not be suitable for all the bioremediation tools and allies. According to Starhawk, permaculture teacher, designer and founder of Earth Activist Training, "If you have a sunny spot, sunflowers make sense. If you have a shady backyard, you could do mushrooms. But you are not going to be able to grow sunflowers in the shade."[1] Sometimes bacteria work really well on a contaminant or under certain conditions, and sometimes fungi do. It all depends. Figure out which tools and allies it makes sense to work with, depending on what conditions you have on-site and what materials you have on hand.

## 6. Engage Community, Neighbors and Necessary Allies

If your site is in a residential or publicly used area, let folks know about what you will be doing. Ask for their input, their support and their participation. Contact local community groups and businesses, put up posters, drop off pamphlets, surveys or letters for neighbors, go door-to-door and host conversations with folks. Put on an info night; organize a community design charette or brainstorming event, as well as community work parties. The more community investment you have, the better chance that you will have extra hands for the project, as well as political support. You may find amazing local expertise. You will also have a better chance of the community wanting to take care of the site, protecting it from redevelopment or vandalism. Also, take the time to contact potential allies who could help with the project: university professors with knowledge on soils and contamination and access to testing facilities, local supportive council members, neighborhood associations, permaculture or gardening groups, mushroom farms and greenhouse operators are just a few folks who could help you with expertise and material.

## 7. Find Funding

Though community scale bioremediation and restoration are far cheaper than the more conventional industrial approaches to site cleanup, you will still require some funding to get the job done. You will need to purchase materials like tools, seeds, plant starts, mushroom spawn, irrigation, materials for fencing and signage — and soil, water and plant testing. If you are jumping through government hoops, they may require you to enlist the services of professionals to approve of your plans or check your work. These costs can add up, so how are you going to cover it all? Funding can come in many forms. You can write grants to different foundations or government programs, or you can approach certain companies who may be feeling guilty and want to either legitimately help out or at least greenwash their toxic underbelly with a sweet, down-home community project. Students at a university can apply to student group funding sources for the project. You can approach local businesses, faith groups or wealthy folks who've got a hankering for philanthropy and community eco-revival. You can also put on fundraisers, bake sales, concerts, go door-to-door and engage your neighborhood in gathering grassroots funding to invest in the project. There are lots of options out there, but the point is you are going to need some money to rock this work. How resourceful you are at getting in-kind donations (such as seeds, plant starts, tools, irrigation, greenhouse material, rain barrels) will reduce the amount of cash funding you need. Regardless, take the time to get enough basic funding for the project you will be doing and consider how you will be able to cover the costs of the project not just in its first few months, but over several years.

## 8. Figure Out How to Get the Materials Necessary for Earth Repair

You may be making/growing your own microbes, plants and fungal allies, or you may be contacting a seed company, greenhouse or mushroom spawn company. If your project requires you to make significant amounts of compost, how will you do this? From where will you get the materials to build the most amazing, microbe-laden, medicinal compost possible? What tools and equipment will you need to work with the land?

How will you irrigate your site? If you are engaging with phytoremediation especially, irrigation is key; a poorly irrigated site can spell the end of all your hard work. It is worth investing in good irrigation and a low-energy, efficient system that will keep your plants alive and thriving. Good irrigation setups can also conserve water more than cheaper sprinklers. If you have committed human power or a moist climate, irrigation may be less of a concern as you don't need fancy irrigation if you have enough hands that are willing to water using hoses, sprinklers and watering cans. Also consider harvesting rainwater by installing rain barrels on-site if there is a structure where clean water can be collected safely via its roof.

Another way to keep your site hydrated, especially in really dry climates, is using greywater. Greywater refers to the wastewater collected from things like your bath, kitchen sink and laundry. It should only be used if you are certain that all the soaps and detergents being used are biodegradable and that no harmful chemicals are present. This water can then be directed to your site either directly or with some filtration to water plants and trees, or it can fuel a constructed wetland.

## 9. Initial Site Cleanup: Getting Rid of Debris

Some sites will need physical removal of debris before you can start biologically preparing it for micro, myco or phytoremediation. Sometimes industry leaves behind garbage and old equipment, other times people take a vacant lot for a dumping ground and throw all sorts of weird garbage there. Cleaning up garbage and removing debris is a hard and not very fun part of site prep, but it's very important. It's also critical that you take care with debris removal, wearing thick gloves, proper footwear, heavier pants and clothes to make sure you don't get cut up or get strange substances on your skin. Wearing some sort of face mask may also be appropriate depending on the situation. If you have any concern about strange smelling or looking liquids or solids on a site, either spilled or in barrels, do not move them yourself. Don't touch them and don't inhale the odors deeply. If you can, bring in the department of environment or appropriate regulatory agency to identify the toxins. Then get them or the company (if they are still around) to dispose of them. You may have to politically organize and pressure them to do it, but it is their responsibility and job, so hold them to it.

## Protective Gear for Earth Repair

It is important to protect yourself to the best of your ability from toxic chemicals and metals by reducing exposure and wearing protective gear. The amount of protection you will need depends on the situation. Suiting up for the garden will likely be a lighter matter than oil spill cleanup. In some situations, gloves, an apron or full length clothing and a face mask may be enough, while others may demand rain gear, Tyvek® coveralls, respirator, eye goggles, boots and more.

### Masks

When it comes to protecting yourself from inhaling heavy metals in the soil dust of your garden or toxic chemicals and fumes from contaminated land or a spill, there is a range of options from cloth or paper masks to industrial-grade respirators. A cloth or paper mask will help keep out some dust or contaminated soil particles, but it will not prevent you from breathing in chemical fumes, as fumes can pass through paper and cloth and leak in around the edges of a loose-fitting mask. A tight-fitting paper mask will protect from dust or contaminated soil particles better than a ☞

loose mask. The mask should conform to your face all the way around. A plastic dust mask will protect from dust and contaminated soil particles better than a loose cloth or tight paper mask, but it will also not stop you from breathing in chemical fumes. To increase your protection against toxic fumes, a rubber respirator is the way to go. A respirator is a more solid mask with filters that may keep you from breathing in chemical fumes. Its protection against fumes is not 100 percent, and protection varies depending on the type of chemicals, the mask's condition, how it is being worn and certain environmental conditions that can cause it to clog up. If your respirator doesn't fit your face properly, contaminated air can leak into your respirator face piece, and you could breathe in hazardous substances. Respirators can be hot and rather uncomfortable to wear, but can protect you from inhaling damaging toxic fumes if properly used. If you are working in a situation where there is serious toxicity, such as an oil spill, make sure you wear a respirator fitted with filters for the specific chemicals there. If you are working in a less toxic situation and don't have a respirator or face mask, you can use a bandana or scarf. But be advised that chemicals will stick to a wet or sweaty scarf or bandana, which makes it more dangerous to use these than to have no mouth protection at all. If you do use a scarf or bandana, rinse and dry it often and know that it does not offer much protection.

## Coverings

If you do not have protective gear like a Tyvek® or Tychem® suit, wear rain gear or a thick canvas or rubber apron if you are in a more toxic situation and full-length clothing for a less toxic situation. See Chapter 9 for more information on protective gear for oil spills. Definitely wear close toed shoes and waterproof gloves if you can. If exposed to chemicals, wash your skin right away with cool or warm water and soap. Avoid hot water because it opens the pores. Also wash your hair. Change and dispose of your clothes and shoes outside of your house or place of work, and if you are washing them or spraying them down, please do so in a properly ventilated room or outside. If you are doing this outside, be sure to watch for runoff and position yourself upstream/upwind of the contaminated site so that you do not spread contamination to a clean area.

## 10. Hit the Ground and Start Bioremediating

Roll up your sleeves and get ready for many months or years of earth repair exploration and dynamic community interaction. And remember to pace yourselves. In the words of one of the founders of the Amazon Mycorenewal Project and professional bioremediator Robert Rawson, "We started off running, trying to get there in a real rapid way. And then we realized that enthusiasm has to shift to more long-term endurance. Five years is nothing to getting something done."[2]

## Art Interventions!

If you are not dealing with toxic and dangerous debris, you may want to repurpose it into some useful form or artwork on the site. Invite carpenters, welders, artists or art students into the mix for this one. Maybe some interesting statues, benches, tables, signs, fencing, containers or installations can be made that can engage and educate folks about the fertility found in all forms of reclamation and regeneration. Creating beauty from debris reduces the volume going to landfills and serves as a reminder of the history of the site as well.

## How Can We Have More Bioremediation Projects in Our Communities?

1. Team up with local universities to get research projects happening at the local level. These projects allow for testing and affordable expert assistance that can produce results that can be used to later sway government bodies.
2. Most of the earth repair in this book is about remediating and regenerating contaminated landscapes. But maybe you want to start with a vacant and damaged lot, and not a contaminated one. Experiment with earth repair practices to assist in the natural recovery of this site, and make it into a demonstration site for the community. With this project under your belt and proof that your group is committed and able to do great work, set your sights on a more controversial and contaminated site for the next project. ☞

3. Pick a site that will give you access with ease — like a community garden or farm that has some contamination in its soils. Consider an arts center, a churchyard or your own backyard. Basically, start small and with a friendly landholder; that way you avoid companies and governments overly spooked out by liability issues found at larger brownfield sites.

4. Start a training program in your community; train folks in grassroots phyto, micro and mycoremediation. If folks are inspired and have the skills, they will be motivated to find places to practice them and take action.

5. Start small and quietly to get your practice down, as well as your skills. Build up trust and then scale up.

6. Don't become discouraged by bureaucratic or regulatory opposition. Many projects and ideas that we think of as common and celebrated today were originally discounted, dismissed or actively suppressed by the mainstream system. It took folks who were unwilling to give up or take no for an answer to change that reality. Keep pushing, and in time, opportunities will open. Sometimes the block you are facing will shift suddenly due to a political, economic or social shift, and when it does, be ready to rush in and throw down!

# Microbial Remediation

*I really came to feel that the soil bacteria and the fungi are our allies. They want to help us. They are waiting for us to get it together. They are not going to save the world unless we do. But if we do they are eager and willing to help us. They are enormous allies for us, and at the same time they are the simplest, humblest organisms on the planet. Humble comes from the root word humus, which means close to the earth. In a very real sense, they are our ancestors. They are the first life-forms, the original life that figured out the basic processes for sustaining life on the planet: photosynthesis, respiration, fermentation, how to use oxygen to burn food and make energy.*
— Starhawk

MICROBIAL REMEDIATION uses microorganisms to either degrade organic contaminants or to bind heavy metals in more inert and less bioavailable forms. Microorganisms break down contaminants by using them as a food source or metabolizing them with a food source. There are aerobic bacteria, and then there are anaerobic bacteria. *Aerobic* processes require an oxygen source, and the end products typically are carbon dioxide, water and salts. *Anaerobic* processes are conducted in the absence of oxygen, and the end products can include methane, hydrogen gas, sulfides, elemental sulfur and dinitrogen gas. Microbial remediation can be facilitated either by breeding bacteria in high numbers and then introducing them into a contaminated area (via aerobic compost, compost tea or a more anaerobic mix that you can purchase known as Efficient Microorganisms) or by creating the right conditions in affected soil or water so that they become an ideal habitat for bacterial growth to occur.

There are a lot of examples of industrial microbial remediation (which industry calls bioremediation) applied to oil spills, groundwater remediation and other contaminated site cleanups. Both industrial and grassroots microbial remediation engage with the pollutant-binding and transforming power of microorganisms, just in different ways and on different scales. The information I share here is not the industrial way of doing microbial remediation. Some of the conventional bioremediation methods in use can result in adverse impacts on the environment. Also, some of the work is beyond the skills and resources of ordinary folks. For example, groundwater remediation is hard to do without the proper equipment (for drilling, oxygenating and pumping) and hydrological knowledge. If done improperly, it could cause more damage than good.

So what you'll find in this chapter are tools you can safely use in your community to facilitate, amplify, inoculate and support the amazing power of microorganisms to break down and bind up contaminants, as well as to replenish the earth beneath your feet. As with most things in nature, these tools are meant to act in concert with each other, to collaborate, to be connected. There are many paths up the microbial mountain to recovery.

## Microbial Remediation Principles for the Grassroots Remediator

### We are Working with Valuable Allies who are Part of the Soil Food Web.

No foray into microbial remediation can begin without exploring the billions of tiny and powerful organisms that can be found living within the soil beneath our feet, as well as on essentially every living surface of the world. In order to properly engage these microorganisms, we need to understand who they are and how they work. Dirt is not soil. Soil is alive; dirt is dead. To grassroots mycoremediator Ja Schindler, soil is "a living material composed of minerals, organisms and the gasses and water in between. Likened to an organ for the Earth, similar to our skin, it is teeming with an enormous diversity of beings, both macro and microscopic, sharing an array of landscapes, food sources and social roles."[1] A single teaspoon of soil can contain billions of microscopic bacteria, fungi, protozoa and nematodes, as well as earthworms, arthropods and other visible crawling creatures.

It is this diverse microbial life that fuels so many of the processes that transform and transport nutrients and minerals to the web of life. It is these organisms who can bind, immobilize and transform a whole range of contaminants into less harmful substances.

### *Microbial Remediation is the Basis for Regenerating a Damaged Site.*

According to Marika Smith of the Victoria Compost Education Centre, "If you are doing any restoration project, healthy soil is the foundation. If you are doing any kind of replantings, it won't work if the soil complexity and soil chemistry aren't there. You really need healthy soil with a humus layer. It can take a few years to build the soil. Sometimes the soil is all right and you can plant right into it. Other times you need to do complete soil repair."[2] Bacteria and fungi feed plants, and plants feed bacteria. When the plant needs a certain mineral, it offers sugars to the microorganisms that can provide that mineral. In exchange for the sugar food created by plants, as well as the habitat created by their root systems, microorganisms transport and transform into more available forms different nutrients and minerals, as well as increase disease resistance in plants. In the process of feeding on plant materials and each other, these microorganisms also produce hormones and enzymes that plants need, and consume or break down pollutants in the soil.

In conventional remediation, bacteria are often fed a chemical-based fertilizer instead of a plant-based food. However in grassroots microbial remediation, we need to work with plants to provide more sustainable, local and balanced food sources for microorganisms. Also, plants can struggle growing in really damaged and contaminated soil with very minimal microbial life. Engaging in microbial remediation by using highly biologically active compost and compost tea will help plants to grow and enhance their ability to survive in adverse conditions by reconnecting the powerful, and enduring relationship between soil microorganisms and plants.

### The Warmer the Weather and the More Oxygen in Soil or Water, the Quicker the Microbial Remediation.

If your soil is extremely compacted or waterlogged, microbial remediation is more of a challenge. If you are trying to restore soil or water health, you need to find ways to bring oxygen back into your site/system. As compaction occurs, oxygen movement slows, and beneficial aerobic microorganisms go to sleep. Less beneficial and sometimes detrimental anaerobic organisms start to flourish. Therefore, if you want to grow the maximum amount of beneficial bacteria or support your site in doing so, you need to make sure that your compost, your compost tea, your soil or the water you are working in have an adequate supply for oxygen and are sufficiently aerated.

Warmer temperatures give rise to more microbial activity, and colder temperatures slow things down considerably. For example, the degradation rate by oil-eating bacteria of oil spilled in colder northern waters is much slower than that of oil-eating bacteria in warmer southern waters. On a smaller scale, a warmer compost will allow for more microbial activity than a cold one.

### The Best Microbial Remediation is Done when There is a Diversity of Species Present.

When inoculating your site with beneficial bacteria, you want to provide a diverse array of these organisms. The more species and types of microorganisms that are present on your site, the higher the chance that some of them will be able to work with key contaminants, as not all microorganisms will be effective with a specific contaminant. Often, the microorganisms best suited to the job of remediating your site are those found on your site already or in your region. If you can find out who these microorganisms are and feed them a variety of foods they like to eat, then you can work to boost their populations and help spur their natural processes of breaking down or binding contaminants.

### Getting Started with Your Grassroots Microbial Remediation Project

The following tools and methods are relatively easy and effective at introducing beneficial microorganisms to assist with site cleanup, recharge your soil ecology on damaged sites and help kick off healing the land. What are the low-tech and high-impact ways that grass-roots remediators can harness the power of these small but mighty microorganisms for earth repair? Through compost, compost tea, ver-micomposting and applying humic acid, to name a few.

## Method #1: Compost, Compost, Compost

For the grassroots remediator, compost is one of the most direct, easy and accessible ways to inoculate a site with beneficial microorganisms, contaminant-binding humic acid, organic matter and fertility. If made correctly, your compost will be full of a diversity of microorganisms that can break down, transform or bind heavy metals and organic contaminants.

Expensive to buy in large quantities but free and easy to make, *compost* is the natural process and practice that decomposes and transforms organic waste sources that we find in excess all around us. By mixing materials high in nitrogen (also knows as greens) with materials high in carbon (known as browns), we can create the ideal conditions for microorganisms to grow and flourish. Through their metabolic processes and time, materials can be broken down into a life-regenerating organic mixture (*humus*) that is rich in nutrients, minerals and microbial life. The humus and humic acid produced via the composting process can either stabilize or assist in the degradation of contaminants. Humic substances have electrically charged sites on their surfaces that can attract and make certain contaminants inert.[3]

In the following sections, I'll cover how to make amazing micro-orgasmic compost that you can bring onto your site to inoculate contaminated soil. I also describe how to compost some types of contaminated soil.

## Micro-Orgasmic Hot Composting (Thermophilic Composting)

Not all compost is created equally! Kitchen or backyard black bins and city compost often lack enough microorganism quantity and diversity for effective remediation. More commercial composts have nutrients and minerals, but they are not created in ways that really allow aerobic microorganisms to flourish. According to Elaine Ingham, a world-renowned soil biologist and composting guru: "There is no substance on this planet that some microorganism will not be able to chew up. The problem is finding the right organism to chew up the right nasties and giving them the right food to do their job. If we just make really good aerobic compost, we are going to get the full set of organisms that we need to get into that soil, like bacteria, protozoa, mycorrhizal fungi, nematodes and micro-arthropods."[4]

Thermophilic compost piles produce the largest number of beneficial microorganisms as they create and maintain ideal conditions for their rapid growth. These piles differ from other types of composting because attention is placed on temperature, the ratios of nitrogen to carbon materials and the need to maintain good aeration in the pile. This form of composting may seem a bit high maintenance, but your site will thank you for it.

### Use the Right Materials — Nitrogen and Carbon

In order to make great compost that packs a medicinal punch, you need to use the proper ingredients in the right ratio. Mixing together the correct balance of carbon and nitrogen is one of the most crucial factors in breaking down organic matter, building heat and creating the appropriate conditions for the right microorganisms to become activated. *Carbon materials* are usually brown and dry, so these would include dried leaves, wood chips, straw, hay, dried grass clippings, shredded newspaper and cardboard. *Nitrogen materials* are usually wetter and greener (though this is not always the case), and some of these include kitchen scraps, dumpstered vegetables, different types of manure (some manure types are higher in nitrogen than others), coffee grounds, okara (soybean waste), comfrey leaves, garden waste, urine and seaweed.

## Phosphorous

In our standard composting practices, we don't always think about phosphorous. For the sake of microbial remediation wonderment, add this important nutrient to the mix! Phosphorous is an essential ingredient in life and is a fertilizer for bacteria, stimulating their growth. Chicken manure is great to add because it contains phosphorous as well as nitrogen. So do bone meal, fish bones, guano and rock phosphate.

A quick word of praise for wood chips: if you are trying to make a compost pile that is about microorganisms, wood chips are great as they have lots of surface area, create air space and allow your compost pile to hold on to its volume instead of shrinking down. Remember, when sourcing your materials, do your best ensure they are organic and have not been treated with chemicals (in the case of plant wastes, vegetables, hay, grass, straw), antibiotics or hormones (in the case of manures).

Organic waste materials all have varying degrees of carbon or nitrogen, and this is known as their *carbon to nitrogen ratio* (C:N). Microorganisms that digest compost need about 30 parts of carbon (for energy) for every one part of nitrogen (for protein) they consume. If you have too much nitrogen in your mix, your compost will heat up too quickly and the nitrogen will turn to ammonia gas (which you can usually smell). Excess nitrogen means you will end up burning up a lot of the organic goodness you want to retain. You will also lose a lot of your nitrogen as it is released in gas. Having the right carbon to nitrogen balance eliminates foul compost odors and potential complaints from neighbors. Too much nitrogen can also make a compost pile too wet, which produces anaerobic (oxygen deficient) conditions that decrease your ability to grow the beneficial aerobic microorganisms so useful in breaking down organic matter and contaminants. If you add too much carbon materials to your pile, it is too dry and will not heat up sufficiently. A friend of mine described it best by comparing the nitrogen in your compost to a gas pedal and the carbon to a brake. He recommended that it is better to err on the side of more carbon than nitrogen, as it is better to keep the pile from getting too hot, and it is easier to pee on the pile to heat it up than to try to cool it down!

When it comes to remediation, the higher the diversity of microorganisms in your compost, the better. It is also important to note that bacteria prefer foods that are higher in nitrogen, and fungi prefer foods that are higher in carbon. You can also add in some key plant wastes that are high in nutrients and minerals (e.g. yarrow, thistle, bracken fern, plantain, garlic) to increase the richness of your compost. Plants like horsetail are high in silica, buckwheat is high in phosphorous, nettles are high in iron and calcium; seaweed provides a whole host of micronutrients that cannot be found anywhere on land. Furthermore,

Get to know local organic waste sources that you have in abundance, figure out where to find them or make the connections to get them.

| Material | Carbon to Nitrogen Ratio (C:N) |
|---|---|
| **Higher in Carbon** | |
| dried leaves | 35-85:1 |
| straw | 80:1 (general); wheat straw has a high amount of carbon, oat straw is lower. |
| woody plant stems | 700:1 |
| pine needles | 60-110:1 |
| cardboard | 400-563:1 |
| newspaper | 398-852:1 |
| telephone books | 772:1 |
| wood chips | 560:1 (hardwoods) 641: 1 (softwoods) |
| sawdust (weathered two months) | 625:1 |
| sawdust (weathered three years) | 142:1 |
| **Higher in Nitrogen** | |
| food scraps | 15:1 |
| coffee grounds | 20:1 |
| vegetable produce | 19:1 |
| fruit produce | 40:1 |
| apple pomace/mash | 13:1 |
| okara (soybean curd residue — anywhere you get tofu) | 11:1 |
| fish scraps | 3.6:1 |
| brewers' mash/brewery waste | 12-17:1 |
| grass clippings | 12-19:1 |
| alfalfa hay | 18:1 |
| garden waste | 30:1 |
| cow manure | 19:1 |
| horse manure | 22-50:1 |
| chicken litter | 10-15:1 |
| sheep manure | 16:1 |
| pig manure | 14:1 |
| humanure | 5-10:1 |
| seaweed | 19:1 |
| urine | 0.8:1 |

Fig. 4.1: *Carbon to nitrogen ratios (estimates and ranges) for some commonly used compost materials.* CREDIT: LEILA DARWISH[5]

the longer compost sits without disturbance, the more it will move from a bacteria-dominated pile (compost better for annuals and vegetables) to a fungal-dominated pile (compost better for trees and shrubs). For remediation purposes, you should aim for a good mix of the two.

## CONSIDER SURFACE AREA

Microorganisms are small, and so their food should be. The smaller your materials are in terms of chunks, the higher their surface area, the easier they can be ingested by microorganisms, the more air they will let through and the quicker they will break down. You can speed up decomposition of carbon materials by cutting them up into smaller pieces. You can cut materials up using pruners, scissors, wood chippers, leaf shredders or lawnmowers.

## LAYER AND MIX

Once you've collected all your materials and chopped a few of them up to manageable size, you can start to create your pile by layering or mixing them up. Most folks alternate carbon and nitrogen layers that are about four to six inches thick. While layering, every so often mix all your materials together (using a pitchfork) to avoid compaction and get an even distribution so that the carbon and nitrogen are close enough to do their magic. It is best to form the base of your pile with rougher materials like twigs, small branches, wood chips and straw to allow air in from beneath the pile. You may also want to put your stinky, more pest-loved materials (such as food scraps, okara and brewery mash) in the middle of the pile. It is a great idea to add some compost from a past pile or from a friend's pile to act as a microorganism inoculant for your new pile. You can use worm castings from your worm bin/vermicompost if you have those as well. A hand or shovel full will do. As you are mixing and building the layers, add some water to the pile (unless your materials are very wet). It is advisable to finish your pile with a top layer of carbon material, as this minimizes odors and flies. Once you have enough materials and enough mass for your pile, you should cover it with a tarp or lid to keep the rain out and the heat in. Poke a stick in the top of the compost pile with a little bucket on top for the tarp to rest upon, allowing for the compost to breathe.

### Build Volume/Critical Mass

In order to reach the appropriate temperature for your microorganisms, your pile should be no smaller than one cubic yard. Microorganisms like to stay warm — and this size of pile, or larger, ensures that there is enough insulation on the outside of the pile to keep the inside warm, which is where the main activity of microorganism growth and organic matter decomposition takes place. Piles that are too small will not be able to generate the temperatures needed for thermophilic/hot composting, and piles that are too big (larger than four cubic yards) can be hard to manage and turn. Some folks make large windrows of compost instead of a square pile, which is fine, but if you make them too high they can get too hot and anaerobic. Furthermore, when you are building your thermophilic compost, it is best to build your pile up to this volume in one day (three days max) and then to not add anything else unless you need to in order to adjust temperature. Incremental piles (where we keep building them slowly) take too long before they reach the appropriate mass to get hot.

### Aerate

Maintain oxygenated conditions by making sure the pile is not overly wet or waterlogged, properly mixed, and minimizing material that is prone to clumping or compacting (grass clippings, sawdust, okara, brewers mash). Balance those denser materials with rougher materials (plants stalks, wood chips, small twigs). Turning a compost pile every so often will also help bring in oxygen and break up settling and compaction (though you don't want to turn it too much as it will also disturb the microbial communities you are trying to build). Air porosity decreases over time as a compost pile settles and becomes denser. Some folks recommend building compost around a *chimney* (a hollow tube with holes drilled into it that maintain passive airflow).

### Monitor Moisture Content

If your compost pile is too wet, you will produce oxygen-deficient conditions that will breed anaerobic microbes. Too dry compost is no good either as microorganisms need water to survive, and without it they cannot do their work and the pile will never heat up. Water also helps soften organic material. A good compost pile should be as wet as a wrung-out sponge, some say having about 50 percent moisture. When you are layering, you may need to add water if you are working with dry material. If you are working with material that is quite wet to begin with, you will want to mix it with dryer materials if you can. If you have sunny weather, leave the pile uncovered for a little bit to try to dry it out. As said previously, you will also want to cover the pile to keep it from getting rained on.

## TEMPERATURE MAKES MICROBIAL MAGIC (VERY IMPORTANT!)

If you've gotten all the other pieces (materials, oxygen, volume, layering and moisture) correct, your compost will now begin to heat up as carbon and nitrogen interact and microorganisms begin to feed on their food sources, reproduce and warm the compost. Once you have finished building your pile, you should buy a compost thermometer (they are fairly cheap at US$15–30) and use it to monitor the temperature of your pile. Check the thermometer daily or every other day. You are aiming for 131–145°F in order to breed bacteria. Avoid going higher than 140°F if you can. Higher than 145°F will kill some of the beneficial microorganisms you are trying to breed. Over the next two to three weeks, you must monitor the compost temperature closely, raise it to the appropriate level and then turn your compost to either lower the temperature or raise it again. It is a good idea to leave enough space around your compost for two piles, so that you can easily flip your pile back and forth. Read closely the following temperature and turning sequence needed in order to support the maximum growth of your microbial remediators:

### Phase 1

It may take several days for the temperature to reach 131°F. If the pile is having trouble getting to this temperature, it may be that your mixture doesn't have enough nitrogen in it. You may want to pee on your pile to help it out, or add another nitrogen source to it. Once it reaches 131°F, allow the pile to plateau (or let it rise to 140°F). If the temperature goes above 145°F, then you need to flip it or you will breed bad bacteria/pathogens. When pile temperature starts to drop of its own accord, flip your compost, basically taking the outside part of the compost and placing it inside and vice versa. Scrape off the outside layer of the pile and put it aside as this layer will now form the inner core of your compost. Find the hottest parts of the compost pile (which will be towards the middle near the top) and put them on the outside.

## Phase 2

Let temperature rise again to 131–140°F and plateau for a few days. Then the temperature will start to drop off. When it drops back down to 122°F, you want to flip the compost again.

## Phase 3

Repeat: let temperature rise to 131–140°F, plateau and then as it begins to drop, flip it one last time!

## Phase 4

Leave the pile alone after that — no more flipping, adding anything, or need to monitor the temperature. Up to this point, the whole process should have taken anywhere from 16–18 days. When the temperature of the compost pile is the same as that of the ambient temperature, then it is time to let your compost cure, which should take at least two weeks. Bacteria (the primary decomposers of organic material) create heat and dominate the early stages of the compost process, while the fungi arrive later, followed by micro-arthropods and nematodes. Over the next little while, the microorganisms will continue to build up their populations and their community will undergo succession. The longer you leave the pile, the more fungal-dominated it becomes. Ideally you would leave it for one to three months to cure for bacterial pile and longer (three to six months) for fungal pile.

No process is perfect, so just make sure your pile heats up to 131°F at the least and never rises over 145°F. Different sources say that you should turn your pile anywhere from three to five times within the first 16 days. The time allotment for the sequence depends on whether you set your compost pile up right and nothing disruptive happens in the interim (like someone accidently takes compost from your active pile or the tarp gets removed and it gets really wet).

## Soil Biology Tests: A Different Kind of Soil Test

Does your soil or compost have enough bacteria, fungi, protozoa and nematodes? What kind of bacteria does it have and what are the native bacteria to your environment? You don't have to do a *soil biology* test or a full *soil food* web analysis in order to bioremediate your site, but these tests can offer you some important information to guide your work and to also see if the compost and compost tea you are making is actually getting the job done.

Labs like Soil Food Web Oregon, Soil Food Web New York, as well as a few Soil Food Web labs in Canada can do these tests, as can universities. Or, if someone you know can use a small microscope to do the shadowing method of soil analysis, it's very quick and can do the trick.

### Apply the Compost

Use your senses to figure out when your compost is ready to use. If it looks dark and broken down and smells earthy and rich, it should be ready to rumble. The compost will inoculate the site with the powerful force of beneficial microorganisms, such as bacteria and fungi. It will also support your phytoremediating plants, creating dual action site cleanup and regeneration.

Opinions differ on the best way to apply compost in order to get microbial remediation rolling on a site. You can either till it in or lay it on the surface. If you decide to just lay compost on the surface, I would at least take a garden fork and poke holes throughout the site to open and loosen up the earth. By laying your compost on the surface, you will be dealing with less risk of disturbing the site, harming your microorganisms and exposing yourself to contaminants. But not digging or tilling it in may increase the time it will take for remediation; typically, bacteria and fungi need to be carried deeper into the soil by protozoa, nematodes, earthworms and/or micro-arthropods. If these organisms are not present in large amounts and you need to move more quickly, then you will likely have to integrate the microorganisms and the compost into the soil through either tillage (which will harm some of the microorganisms in the compost) or by making *grow holes* throughout the site that you dump your compost into, as well as spreading it on the surface for a good mix of coverage. Finally, adding mulch (such as leaves, straw, grass clippings or wood chips) over your freshly applied compost and the rest of your site will create a hospitable environment for beneficial microorganisms. More carbon-based mulches, like dry leaves and wood chips, favor fungi, while more nitrogen-based mulches, like grass clippings, favor bacteria.

## Community Composting Operations

Every community should have a community composting operation because you never know when you'll be needing compost in abundance. And besides, if we are going to light up our communities with local nourishment, we will at the very least need lots of compost on hand for food production. In the case of compost, you can't have too much of it.

According to Scott Kellogg, author and co-founder of the Rhizome Collective and the Radix Center for Ecological Sustainability: "A good use for a piece of potentially contaminated land is a community composting operation. The land may be unfit for growing food, but suitable for surface compost piles."[6] To local government officials and regulators, as well as the company that may own the site, a community project making surface compost piles may seem less risky than food production or phytoremediation, since surface composting does not involve the risks of exposure when people dig into potentially contaminated ground and ingest potentially contaminated food.

Chatting with Australian permaculture design guru and author David Holmgren revealed a powerful story of a community composting operation. "There is a story I would tell about the CERES City Farm in Melbourne, which is the flagship site of an organization of the city farms and community garden movement in Australia. It was started in 1978," recalled Holmgren. "It is on an old industrial waste site, a rubbish tip among various other things. And the first thing that was done on that site was a very large composting operation, basically for the purposes of soil improvement. There was a huge amount of fresh mineral dust, basalt dust from all the stockpiles of bluestone pavers which was actually one of the industrial waste products left on the site, which you can say acted like a mineral ☞

fertilizer. This waste mineral dust combined with this very large composting operation and it actually inoculated the whole site and began a bioremediation process."[7] Built on what was once a landfill and wasteland, today CERES, the Centre for Education and Research in Environmental Strategies, is a not-for-profit environment and education center, urban farm and permaculture nursery.

By using these sites as initial community composting operations, a portion of the compost that is not returned to the community or the excess compost that remains on-site can then be spread across the site periodically, introducing and reinoculating beneficial microorganisms and the humic acid that helps to degrade and bind contaminants on-site. The frequent addition of compost would also build up the soil fertility of the site again. This is critical because the degraded and compacted soils of most contaminated and damaged sites often prevent phytoremediation projects from taking off. Over time and with successive layers of compost added, the rehabilitation of the site would occur, as clean soil, a dynamic soil food web and fertile organic matter will at the very least build up on the surface, eventually at enough of a depth that can be safely planted.

At this point you can move your composting operation onto another damaged site and transition your current site into a different phase of more active bioremediation (if contaminants are still an issue), or the site may be ready to be turned into a community garden or orchard, urban farm, food forest or park. You can also decide whether you want the community composting operation to take up the entire site and be the primary activity, or you can go for more of a hybrid model, rotating compost production with either movable raised beds for veggie production or planting phytoremediating plants into the compost-laden earth.

You will want to approach your community compost work as professionally as you can, taking care to maintain the compost so that it doesn't produce foul odors and legions of rats to piss off the neighbors. Unlike most bioremediation projects, you can actually use what you are making on such a site within a relatively short time frame, and can either sell some of the compost to sustainably fund the project, exchange it for useful supplies, or share it with your community as a way to build a better relationship with folks and support their earth repair work as well.

## Method #2: Vermicomposting: Using the Power of Worms For Soil Remediation

According to one study, PCB levels in sludge on a SUPERFUND Site were reduced by 80 percent in different vermicomposting bioreactors, although the time required for this level of reduction increased with increasing sludge concentrations that were added.[8] Another earthworm remediation study done in the small agricultural community of Muthia in India also showed promising results. Textile and pharmaceutical companies had historically used the region as a toxic waste site, dumping contaminants such as chromium, lead, iron and zinc. A four-year vermiculture bioremediation project was initiated, using both beneficial microbes and 300,000 worms to eat up and break down the contamination on four acres of contaminated land. In the final phase of the study, maize was planted and tested for contamination. The tests showed only trace levels of heavy metals in the soil and crops.[9]

Worms provide several remediative and regenerative services to a contaminated site. Their movement and tunneling through the soil helps aerate it, which helps with making it more habitable for other beneficial microorganisms. As worms migrate, they disperse the pollutants and reduce contaminant concentrations in a given area. Because worms bioaccumulate heavy metals and organic contaminants and are able to consume up to 50 percent of their own body weight in food

Fig. 4.2: *Worm compost.* CREDIT: VICTORIA COMPOST EDUCATION CENTER

each day, these tiny creatures do really compost down, concentrate and reduce the volume of contaminated soil they are working with. *Worm castings* (their poop) greatly increase the fertility of soil and also inoculate it with beneficial bacteria. The digestive tracts of earthworms contain the bacteria *Pseudomonas fluorescens*, which allows them to digest things like lignin in leaf litter. What folks have been finding with bioremediation (especially with mycoremediation) is that creatures that can digest lignin can usually digest hydrocarbons as well as other organic chemicals. An increased number of bacteria in the system can munch away, break down or bind contaminants.

Another way that worms are useful in remediation is that, as they consume organic matter in the soil, they retain the ingested contaminants (heavy metals) in their body tissues rather than excreting them. This binds the contaminants and makes them less bioavailable, as long as the worm is alive. Supposedly, when those worms die, other worms and microbes ingest the worm proteins so the toxins remain sequestered and aren't released and taken up by plants — unless all the worms in a region died suddenly and the soil was damaged and degraded in such a way that caused microorganisms to go dormant or die as well. This could cause an unfortunate situation where contaminants would be re-released into the soil. In some remediation studies, worms have been collected and incinerated in order to remove the contaminants from the soil, however this would be out of the scope of most grassroots remediators and may have other detrimental environmental impacts. There is also concern that the contaminants being bioaccumulated by the worms can then end up travelling and contaminating other birds and animals in the food chain that feed on the worms.

If you don't have attachment issues with worms and you have some way to keep hungry birds and animals off your site, then you may want to consider integrating vermiculture into your remediation, to enhance your microbial or phytoremediation efforts, especially if you are dealing with low to moderate contamination. Worm castings are incredible fertilizers and will provide rich nourishment to subsequent phytoremediation plantings on your site. Their castings have seven times the phosphate; ten times the available potash; five times the nitrogen; three times the usable magnesium; and one and a half times the calcium of regular soil.[10] Besides their role in remediation, the presence of existing worms in the soil indicates a healthy soil food web, since organic matter, bacteria, fungi, protozoa and nematodes must all be present to support a worm population.

Experiment with simple vermicomposting at home, and if you are really inspired and want to take it up a notch in scale, take a few classes on vermicomposting or intern at a worm farm. Having access to worm castings will help to inoculate both your compost and compost tea with beneficial bacteria at the very least, as well as feed your garden and increase its resilience.

## Binding Contaminants with Humic Acid, Mycorrhizal Fungi and the Soil Food Web: An Interview with Dr. Elaine Ingham

Dr. Elaine Ingham is a world-renowned soil biologist who studies the microbial life of the soil. She founded Soil Foodweb, Inc. in 1996 to assist farmers in growing more resilient crops by understanding and improving soil biology. She is an author, educator, compost and compost tea master and currently the Chief Scientist at the Rodale Institute. When I interviewed her for *Earth Repair*, Dr. Ingham emphasized the importance of doing good detective work to figure out what contamination you had in the soil and where it was coming from, followed by figuring out, using a soil biology test, what microorganism allies you had already present on the site. She also spoke about the ability of *humic acid* to bind contaminants at the soil level, and some of the other microbial ways to take care of contamination. Here are some highlights from my interview with her.

### On Applying Humic Acid to Bind Heavy Metals and Contaminants

Dr. Ingham: Come in with a liquid humic acid (or dry if that's all you got) because that initial application is really good at immediately complexing whatever toxic materials or high salt levels that are in the soil. Basically whatever problem chemical that is in that soil will get tied up immediately on the surfaces of that humic acid. Humic acid is very reactive — it has lots of positive and negative charges all over the surfaces. The thing is that humic acid will only hold those heavy metals or chemicals for about two weeks. After that it releases it back into the soil. So when you apply the humates, you have to come in and apply the biology at the same time. Organic matter in the form of compost is a great way to do that. Because now you are not only putting in more humic acid (as that's a huge component of compost), but you are also adding the organisms that can work on the contaminants bound up by the humic acid and further bind and transform these metals and chemicals so that you need not worry about the toxics being released any time in the near future.

With your humic acid, compost and all your organisms, you have provided exactly the right habitat and conditions for all those toxic chemicals to be decomposed and to be put into the structure, the organic matter, in a way that they will never be released in a high enough concentration to do harm. As long as you are putting the system together and letting it be sustainable (no tilling, no disturbance), you allow for successional processes to take place and build that organic matter. And the best plant growth medium is pure 100 percent organic compost, as long it is aerated compost. The compost has to be aerobic, not anaerobic!

If you want humic acid just to complex things, you can use a Leonardite-derived humic acid. But you wouldn't use this for a compost tea because it's been denatured and it needs time to be renatured by biology and you don't have enough time with a tea brew to do that. Plus a Leonardite-derived humic acid is expensive, around the range of US$50 per gallon. What is the easiest place to get good humic acid? From your compost!

## On Working with Mycorrhizal Fungi

Dr. Ingham: Great for binding contaminants and keeping them from moving up into the plants, the mycorrhizal fungi will take that heavy metal in the soil and pull it into the hyphae and lay it down on the outside of the roots of the plant and block it from moving into the plant or through the soil. Now that plant can get growing, and you can eat the apple and the pears and fruit of the plant because the mycorrhizal fungi will not let the heavy metal into that plant! Some research done on the Chernobyl nuclear disaster showed that the only foods that didn't have radioactive materials were those grown in organic gardens that had good levels of mycorrhizal fungi on the root system and strong soil biology.

When you add your compost to your site, you can add the spores of mycorrhizal fungi into it. If you are using compost tea, you would add the spores to the tea or your compost extract. You only put mycorrhizal fungi in if you will be putting the compost or tea into the root zone, so basically use it when you are planting. The easiest ways to get mycorrhizal spores onto a plant is to soak the seed in a compost tea where you added the mycorrhizal spores just before you add the seeds. Those mycorrhizal spores will glom onto the seeds so fast. Put a teaspoon of the inoculum into those five gallons of compost tea and seeds and let soak for five minutes before pulling out seeds (say you are soaking 150 pounds of seed). When you use mycorrhizal fungi this way or in your compost, you end up using a lot less of it and the quantity stretches farther than if you apply it the way the instructions on the product tell you to. The instructions often tell you to put a bunch of the spores in water and spray it out on the earth, but that is not the best or most efficient way to do it as it doesn't get it to the root system necessarily. Treating the seeds is the best way. Another good option is to dip the roots of transplants or trees/shrubs into a mix of compost tea and mycorrhizal spores before planting them.[11]

## Method #3: Engaging the Contaminant-Binding Power of Mycorrhizal Fungi

*Mycorrhizae* are fungi that have symbiotic associations with plant roots. Plant roots provide these fungi with food (carbohydrates) via exudates and in return the mycorrhizal fungi form intricate dynamic webs that bring nutrients and water to the plant, increasing its health and resilience. Ninety to ninety-five percent of plants form mycorrhizal relationships. In disturbed, damaged and compacted soils, mycorrhizal fungi may be absent. Unlike the other microorganisms that have been mentioned in this chapter, these critical fungi cannot be grown or multiplied in compost or compost tea.

You can purchase mycorrhizal inoculum from specialty nurseries or places that specialize in soil amendments and compost teas. There are also two types of mycorrhizal fungi — endomycorrhizal and ectomycorrhizal. Most plants prefer endomycorrhizal, but there are some very common plants that prefer ectomycorrhizal. If you are following

Dr. Elaine Ingham's advice to add the spores to compost tea, be sure to add them once the tea has finished brewing, and apply the mixture as soon as you can.

If you are feeling wild and adventuresome, you could try to find some common visible fungi that form mycorrhizal relationships with different plant species in your area and use them as an inoculum. Some common and edible ones to look for would be chanterelle, king bolete or shrimp russula. Also *Gomphus* (pig's ears), *Hydnellum* (hedgehog and hawk's wing types), truffles (*Tuber* spp.) and *Scleroderma* (earth balls) are species that can be used.

"For gilled and toothed mushrooms, put fresh specimens upright in a bucket of rainwater overnight. The spores will release into the water and can then be poured into the newer root zone of the plant, or over grassy area or ground," explained grassroots mycoremediator Ja Schindler. "You can also dry mushrooms for later applications, as there is no real time limit assumed for dormant (dry) spore life. For truffles and earth balls, just crumble them into a powder, dump into water, then apply." Another method that Schindler uses is to simply collect the mushrooms, place them with the cap up on ground you want to inoculate and then cover with a few inches of leaves.[12] This method could be particularly useful if you match these species up to the plants they tend to support in the wild.

## Method #4: Actively Aerated Compost Tea

According to the *Toolbox for Sustainable City Living*, actively aerated compost tea is a "water-based oxygen rich culture containing large populations of beneficial aerobic bacteria, nematodes, fungi and protozoa, which can be used to bioremediate toxins."[13] Good compost tea should contain thousands of beneficial microorganisms; this increases the chances that some of them will be able to bind and break down the range of contaminants on your site. Compost tea allows you to amplify a small amount of compost into a dispersible liquid form, helping a little compost go a lot farther.

Compost tea is relatively easy, cheap and fun to make — it is also a really great activity to do with kids. It requires an inoculant of beneficial bacteria and fungi, some key food sources, dechlorinated water, oxygen and agitation.

# Compost Tea Recipe

This is a simple and standard recipe for five gallons of compost tea. The proportions can be multiplied for larger batches.

## Ingredients and Supplies

- five-gallon bucket (make sure it is clean!)
- water (either rainwater, pond or if tap — leave it sitting and uncovered for 24 hours to off-gas any chlorine)
- 1 cup of inoculant (worm castings or aerobic compost)
- ¼ cup of food: comprised of equal parts of unsulphured molasses, fish hydrolase, kelp and 1 tablespoon of humic acid
- 1 compost tea bag/stocking
- air pump
- plastic watering can or backpack sprayer (one that has never been used for chemical applications)

# How to Make Compost Tea

1. Pretreat your compost to increase its inoculant and fungal power. Take your compost inoculant and add some humic acid, fish hydrolase, or soaked oats to it. Put it into a shallow tray and mix it up well. Then let it sit for two to three days. This encourages fresh microorganism growth in the tea. Many recipes don't require you to pretreat your compost: you can treat this as an optional step or you can see it as a way to increase the effectiveness of your brew.

2. Fill a bucket with non-chlorinated water. Water temperature is ideally 55–80°F.

3. Put the airstone in the bottom of the bucket, attach the air pump and let it start to bubble. Make sure there is enough oxygen and agitation moving through your liquid; if not, get a more powerful pump or move to the gang valve and three-bubbler approach. Remember, you are looking for more of a churning or rolling boil, not simply fine bubbles. ☞

4. Put inoculant and food in the stocking or mesh bag, tie off the end and suspend it in the water.

5. Let the whole brew bubble for 24 hours and for no longer than 36 hours. After 36 hours, if the tea received insufficient oxygen or too much food, anaerobic organisms will overcome the beneficial aerobic organisms. It will be obvious if the tea went anaerobic, because it will stink! If that has happened, pour it out away from garden plants and start over. One thing to be aware of — just because your compost tea smells earthy and sweet (which is the smell you are going for), it does not mean that it packs a microbial punch, as that smell can also come from molasses. If possible, do a soil biology test of your first few batches to see if you truly are rocking the microbe production.

6. Pour the mixture through a strainer to remove large debris so that it doesn't clog your backpack sprayer or plastic watering can (supposedly bacteria can react with some metal cans).

7. Make sure to clean your bucket and pump for your next round of tea. Use a non-toxic, environmentally friendly, biodegradable cleaner.

## Inoculant

Worm castings and aerobic compost are the best inoculant choices. Worm castings are a great inoculant because worms use bacteria instead of digestive acids in their stomachs to break down food. The castings are rich in beneficial microorganisms, some of which have been found to be effective in breaking down certain contaminants. Worm castings are also a source of humic acid, which is a good food source for your tea. Similarly, good aerobic compost (especially thermophilic compost) is a great inoculant; if made properly, it should be full of beneficial microorganisms.

The quality of the compost used to make compost tea is really critical. The tea can only amplify the biology already present in the compost. So you want an incredibly biologically active compost, ideally one that has at least both bacteria and fungi, to serve as your inoculant. Compost piles that have been curing for three to six months are more likely to be fungal-dominated, while piles that have been curing for one to three months, tend to be more bacteria-dominated. I've heard several compost tea experts say that it can be difficult to get a good amount of fungi. If you can use a more fungal-dominated compost pile as an inoculant, that could give you a bit of an advantage. If you also happen to be cultivating mushrooms, you could try adding spent spawn to your compost to increase its fungal load.

Bacteria are very easy to grow in your tea — they are easy to extract and they like growing in the water. If you test your tea and find it to be fungi deficient but strong in bacteria, it is still good for inoculating your site with beneficial bacteria. You will just have to find another way to get replenish the fungi in your soil food web.

## Food Sources

The food sources you add to your compost tea will determine the composition of microorganisms that grow in it, as bacteria and fungi favor different food sources. Different recipes I found called for different ingredients, and these different ingredients allow you to select for a more bacterial or fungal tea. A mixture of these foods will create a tea with both bacteria and fungi, which is ideal for the remediation of contaminants.

Food sources for bacteria include simple sugars, simple proteins and simple carbohydrates. The most commonly used food source in compost tea recipes seems to be unsulphured molasses. Some other bacteria food sources include fruit juice, cane syrup and fish emulsion. Food for fungi include complex sugars, amino sugars and complex proteins. The most commonly used compost tea food sources for fungi are fish hydrolysate, kelp/seaweed and humic acid. Some additional food sources include fulvic acids, soybean meal, oat bran, oatmeal, fish oils, cellulose, lignin, cutins, rock phosphate dust, fruit pulp (oranges, apples and blueberries) and aloe vera extract (without preservatives). The more types of food added, the greater the diversity of species of microorganisms likely to be present.

Some grassroots compost tea brewers I spoke with preferred to avoid purchasing commercial products (like humic acid) for making their tea, and instead felt that it was sufficient to use their compost and worm castings as an inoculant, along with some dynamic accumulating and nutrient/mineral rich plants, weeds and seaweeds. If you want to go this route, I suggest you get your tea tested to make sure it is indeed productive in beneficial microorganisms.

Fig. 4.3: *Compost tea.* CREDIT: STARHAWK

## YOUR COMPOST TEA BAG OR STOCKING

Many grassroots compost tea brewers I know use a nylon stocking to hold inoculant. The mesh should be at least 400 micrometers to allow fungi and nematodes to flow through. However, some compost tea brewers claim that nylon is not the best material to use, and recommend using a non-sticky compost bag (like a polyester mesh screen) which will allow for more fungal extraction.[14] For optimal extraction, it is also important that you put your inoculant and foods in the bag (see p. 57) and not just directly in the water.

## AERATION AND AGITATION

Use an air pump/bubbler to keep your tea aerobic. Some of the sources I found indicated that an aquarium pump connected to an airstone could supply enough aeration for a five-gallon batch of compost tea. Other sources suggest you use an air pump, some tubing, a gang valve and three bubblers to make sure the entire tea gets aerated.

What many folks don't seem to know is that you need both aeration and agitation for effective compost tea brewing. Lots of compost tea brewers have pumps or bubblers that provide good aeration, but they may not provide the necessary agitation you need to truly aerate the water and knock the organisms, like fungi, off the organic matter and into solution. Fine bubbles don't aerate water. It's the breaking of the surface of the water that gets oxygen into it. So instead of a lightly bubbling compost tea, you should aim for more of a rolling boil, or churning. To achieve this, you may have to play around with a few different aquarium pumps or generative blowers. Some sources suggested using a high-pressure (3.9 psi), high-volume air pump (17 gallons per minute). Avoid using air compressors as they can damage microorganisms.

Remember, these pumps need a power source, and the tea needs to be aerated constantly — so make sure no one turns off the pump at night. In a post-disaster situation where power may be more difficult to come by — or if you live somewhere where electricity is a touchy thing — it may be harder to make compost tea properly.

## Applying Compost Tea

Use your compost tea within four hours of turning off the bubbler, since after that amount of time without oxygen your aerobic microorganisms will begin to die. At this point, you can bring the tea to your site and apply it directly onto the contaminated and/or damaged land, a spill area or onto your phytoremediating plants to increase their health. It is best to apply your tea to moist soil or after a rain, on a cloudy morning or in the evening as some microorganisms do not like baking in the hot sun. If you are applying your tea with a sprayer, make sure that the sprayer doesn't need too high a pressure and that the velocity of the spray is slow — the microbes you are working with benefit from gentle treatment. If you will take more than four hours to do all your spraying, then consider a way to bring your pump with you and hook it up to keep the solution aerated. You can also take a digging fork or piece of rebar and make holes throughout your site to loosen soil and give the microorganisms a way to move more rapidly down to where the contamination may be. Some sources recommend that you use a minimum of about one gallon of tea for 1,000 square feet of contaminated land. When you are using tea for remediation (drenching the soil instead of spraying plants) you do not need to dilute your tea. Finally, apply tea several times, waiting anywhere from two weeks to one month between applications.

If you are going to be making large brews of compost tea, you can use a rain barrel-sized container. Just make sure to adjust the proportions of inoculant and food and get a strong enough pump or two to ensure the barrel is properly oxygenated and agitated. You can also purchase pre-made compost tea brewer systems that ranges from a five-gallon system priced around US$180 to a 1,000-gallon compost tea brewer for US$7,000.

## EM (Efficient Microorganisms)

Efficient Microorganisms is a ready-made product which has a long shelf-life because it works with anaerobic bacteria. Efficient Microorganisms contains only between 8 and 12 different microorganisms unlike the potential thousands found in compost tea, but its strength is in how these key microorganisms work together and amplify each other's work and effectiveness. EM may also do better in waterlogged settings, and you can store it on a shelf. The basic groups of microorganisms in EM are lactic acid bacteria (commonly found in yogurt and cheeses), yeast (found in bread and beer) and phototrophic bacteria (related to blue-green algae). EM is produced through natural fermentation processes. The product was originally developed to increase crop yields, but the microorganisms in EM have been found to produce bioactive substances, vitamins, hormones, enzymes, amino acids and antibiotics which also can enrich and potentially detoxify the soil. You can buy it online or from specialty nurseries.

## Professional Bioremediation and Composting Oil-Contaminated Soils

Robert Rawson is one of the founders of the Amazon Mycorenewal Project. He is also the General Manager for the Graton Community Services District and has been a wastewater management operator for over 36 years. The president of two bioremediation companies (International Wastewater Solutions Corporation and the Pseudonym Corporation), Rawson is a professional bioremediation expert with experience growing and deploying non-toxic bacterial and bioaugmentation products that are appropriate for petroleum spills on land and in various aquatic environments.

## Compost Tea Goes Pro

Though many of the foundations (like compost tea and composting) are the same as those I've already outlined in this chapter, Rawson's work takes it up a notch scientifically, and this is where we enter into the territory of professional bioremediation. When it comes to treating oil-contaminated soils, which Rawson has done with the Amazon Mycorenewal Project in Ecuador as well as in Egypt, Mexico and the USA, he harnesses the remediating power of some very specific bacteria. He has made them into a formula and patented it, calling it IOS 500.

Using these bacterial allies, he brews some pretty refined and large batches of bacterial solution and applies them either to wastewater or to soils contaminated with oil and other substances. His team then works to compost the soils down, usually by windrow composting. Unfortunately, he couldn't tell me how he isolates these bacteria or makes his formula, as that's how he makes his living. I asked him whether the aerated compost tea made by most grassroots bioremediators and garden lovers could be equally as effective in remediating oil contaminated soil. "I love compost tea — it's a great thing. Compost tea will contain most if not all the strains of bacteria I am talking about, as well as others. So you will have a material that will most likely work and that is very inexpensive," he replied. "But I don't take that kind of risk because time is money on big projects. If compost tea is missing one of the key bacteria, it might not break down one of the target contaminants you are working on. You can stimulate local soil bacteria, but professionals cannot necessarily depend on them in sufficient numbers. Also, if local bacteria were alone effective at consuming carbon, then we have to ask why does petroleum contamination remain unchanged in some places for 50 years or more? It's a matter of putting in the right bacteria to augment native microorganisms, and also creating the right environmental conditions in place, such as temperature, food, pH, oxygen. That is probably even more important than the right bacteria: if you create the right environment, they will come, maybe not in sufficient numbers, but they will come."[15]

Rawson shared with me that the original inventor of the IOS 500 formula started off by simply making compost tea but had not yet *refined* his recipe. Since then, Rawson and his company have refined the organisms. "There is a difference between saying that the bacteria in the gut of an earthworm is a *Pseudomonas*, versus taking a *Pseudomonas* out of an industrial waste treatment facility that has been breaking down phenolic compounds, and then building that culture up and cloning that specific culture. So that it is known to have the specific genes and enzymes that would make it able to do the job that you really want. You could take a similar related culture and put it in the same environment and it may or may not perform as well because it may or may not have that gene, or that gene may not have been activated. You need to have the right organism and provide it with the right environmental stimulus in order for it to activate so that it can go do what it has to do," added Rawson.[16]

### Windrow Composting of Oil-Contaminated Soil

When asked how he handles oil contaminated soil, Rawson replied, "I default to basics, simple basics are the things to use. For example, we've windrow-composted 50-year-old volatized crude oil, high asphalptene content, in swamps in Mexico. We've broken old crude oil-contaminated soil down in 45 days by composting it in a windrow with our bacteria. We use manure and dirt, and we lasagna that. We turn it after the heat comes up. We apply our bacteria in each turn, we use them to regulate temperature and stimulate the enzymatic process."[17]

Earlier in the chapter, I described how to make thermophilic compost to inoculate your site with the power of beneficial bacteria. There is also another way to use the process of thermophilic composting in remediation, which is to basically compost the contamination itself. When dealing with oil-contaminated soil, something as simple as composting the contaminated earth, with the help of petroleum-eating bacteria, has been proven successful by professionals like Rawson. Windrow composting, which refers to compost being laid out in long high rows instead of a static pile, seems to be the  preferred method of professional bioremediators. It is easy to make a windrow using a front-end loader, which means you can work with a lot of soil and you can also avoid having many people exposed to the contaminants.

## Tips for Windrow Composting Oil-Contaminated Soil

You need to blend the right ratios of raw materials in the windrow. Because your soil is contaminated with hydrocarbons, you are dealing with a lot of carbon there, but the carbon is not in a very digestible form. So besides your contaminated soil, you will be looking to layer it with materials high in nitrogen and available carbon. Rawson recommends starting off with a thick layer of manure (usually chicken manure or cow manure, or some other "green"/high in nitrogen material). He recommends chicken manure because it is a source of phosphorous, which is a fertilizer for bacteria and stimulates their growth. Basically, you are aiming for a 3:1 ratio of contaminated soil to manure. Wood chips and straw can also be added to balance nutrients and serve as a bulking agent.

After that, deposit a layer of your contaminated soil and then alternate that with another layer of manure and build them up until you get significant mass. At this point, some folks mix everything together (preferable with a front-end loader if you have one, which you would want if you are dealing with a large volume of soil) to avoid compaction, however this is not advisable if you are dealing with soil that has a high crude oil content because it could become an unmanageable mess. Once you have built up your layers, then cap your windrow compost with a layer of manure. When Rawson builds his windrows, he sprays his bacterial solution on each layer to kick-start the bioremediation process as well as help with moisture content. Instead of his special solution, you could spray compost tea. Finally, your windrow should be between 4 and 9 feet high and up to 12 feet across.

All the factors mentioned in the composting basics section earlier in this chapter apply — in terms of moisture content, temperature and oxygen. As windrow composts can get incredibly hot, monitor temperature in the middle of the pile using a compost thermometer. For windrow composting to work, the row needs to be mixed regularly (according to temperature peaks) to maintain aeration. You must turn the pile to mix the different zones of a windrow to allow for oxygen flow and equal composting of the contaminants to occur. The middle of a compost pile quickly becomes anaerobic, while the active portion of the pile does most of the work.

Composting a large amount of petroleum-contaminated soil may be out of the realm of the grassroots bioremediator, so you may want to approach the professionals on this one and see if they will help you out pro-bono, for a reduced rate or at least give you some tips. It is also important to note that several types of asphalt only melt at a temperature far higher than bacteria in a compost pile can stand. If you decide to go the route of windrow composting contaminated soil, be sure to take the proper precautions and to test your compost after to be sure that the process was effective.

Finally, if you are working with a smaller amount of petroleum-contaminated soil, consider working with batch/pile compost instead, as it will be easier to handle manually. Also, this is toxic stuff (crude, PAHs, VOCs), so you will likely want to wear protective gear so as not to inhale fumes, and you will want to make sure your compost can't leach any of its toxic contents into clean soil or contaminate surface or subsurface water. This must also be taken into consideration when windrow composting. It's a good idea to locate your pile or row in a stable place where runoff and leaching can be contained; also put down a thick layer of manure or "green" at the base of your pile. If you are working on an already contaminated site and doing these piles *in situ*, you can't really do more harm than is already present, but if the oil-contaminated dirt is being moved off-site for composting, use appropriate protocol and precautions to avoid spreading contamination to clean environments. In most cases you will have to follow the guidelines of a local regulatory authority.

## Putting It All Together — Using Microbes and Plants for Urban Remediation in Chicago: An Interview with Nance Klehm

Nance Klehm is an ecological systems designer and bioinstigator based in Chicago. Like many old industrial cities, Chicago has manufacturing plants right within the city limits, and as a result its neighborhoods, soils and waterways have a buildup of contamination. It has been said that it is unsafe to grow food anywhere in Chicago unless you are willing to import clean soil. "Conventional remediation is all about removing the first two feet of soil, replacing it with crushed limestone and then capping it off with a frosting of topsoil," explained Klehm. "There is a brownfield that was remediated by the USA EPA just around the corner from where I live. It's about an acre and a half. There is an environmental justice group that finally got the go-ahead to use it as a growing space and small market area. But the amount of soil they would need to bring in to cover the site would cost almost US$30,000. Most operations that I know — if they are really small-scale or neighborhood-based — just ignore that expensive approach and grow directly in the soil. Or they do raised beds and they do it directly above cement. Or they will do a lasagna/sheet mulch approach with cardboard and wood chips up at least a foot and a half and then start loading up compost and growing on top, just waiting for the mulch to break down and build soil."

To Klehm, you cannot remediate a site with plants unless you start first with the soil biology. "Soils can be so compromised and compacted. To even get them to the point where they are a good growing base for plants means I need to work with the microbes first," Klehm said. "If I can get a soil in balance, so that it is really microbially active — so that it has the nematodes, the protozoa, the bacteria, micro-arthropods and the fungi — then that is the baseline. I usually spend a lot of time on that before I get into the plants."

Describing her approach to remediation, Klehm had the following advice to offer: "I'd start with a baseline soil test, and then I would probably come in with the compost, the compost teas and plant extracts on that site first for a year and then test again. Then I would introduce a really diverse community of plants. And continue to feed the soil with microbes. Often in a site, we have to aerate the soil in some way. I spend a lot of time doing that before I even introduce plants." On sites where there is significant soil compaction (which leads to anaerobic soil conditions) Klehm has used rebar to poke holes in the ground to help get more airflow into the soil. This creates a more aerobic environment, allowing for microbes from compost and compost tea to make their way down into the soil and remediate at a deeper level. "When I introduce plants, I usually introduce woody plants because they are tough, they have incredible root masses and they will anchor a site. I work with woody plants first and then other perennials. Then I go back to those annuals that you can pull out and dispose of," continued Klehm.

When it comes to using plants for remediation, Klehm likes using plants that create an active rhizosphere in the soil and that stabilize the site and provide habitat for beneficial microorganisms. The *rhizosphere* is the area of interaction between the surface of the plant root and the area surrounding it, which is filled with bacteria, other microorganisms and soil debris.[18] She also selects plants for the site that will produce lots of food for an active soil food web. Often when we see vacant lots or disturbed sites, they are covered with weeds and other pioneering species. What is interesting about weeds is that though they can re-plant a site quickly and carry out some key functions, when compared to other plants they actual pump out very little food (in the form of exudates) from their roots for the soil microbes. This is because weeds are trying to move through their life cycle quickly in order to further spread onto the site. "They are only sharing like 20 percent of their sugars with bacteria and fungi, whereas a tree is sharing 60 percent. Trees also grow an incredible root mass," said Klehm. "I'll work with perennial grasses, a lot of things that are considered cover crops, nitrogen fixers or green manures. But I also try to get a couple of trees in there because they will really send the roots far and wide and create a lot of sugar exchange, calling forth the fungi and bacteria." When working with annuals and cover crops, Klehm views them in the following way: "I think of them as stabilizing the soil from erosion, creating an aerobic situation and making it ready for the plants that I want to use later, in two or three years. For me, remediation of a site takes at least five years or longer."

On a neighborhood site that Klehm is transforming from a desiccated vacant lot to nourishing gardens, they have started with microbial remediation. "We began by building very large berms using the technique of layering carbon and nitrogen in them, and they are set to compost down and build clean soil. They are maybe three feet wide and two feet tall," said Klehm. The berms will act as raised beds, so that plants can be safely grown in them on a quicker time line than it would take to remediate what is likely contaminated soil below. Following that, Klehm and her work party added cardboard and mulch to create pathways on the site.

Many urban gardeners and farmers stop at this point — they import or build soil up on a site, mulch the rest and move on to growing their veggies. However, where they stop is where much of Klehm's work begins. "We built beds up but we also addressed down into the soil throughout the site. We cored into the ground underneath the paths and through the beds, and we put big cardboard tubes down that we had punched holes in. We filled them with compost to do a soil biology drench. The reason is that we are trying to use the biology of the site to unleash some of the minerals in the clay below. These cardboard tubes will stick up above the mulch. We can test the soil around them to see how we heal things over time. Then we planted a lot of trees and a lot of woody perennials. We are going to use the rhizospheres and the woody root masses to get the biology going. And once the fungal and bacterial communities are really strong, they will start helping to bind some of those contaminants, while also building nutrients through the site," explained Klehm. "It is not just our legacy to grow food the way we want to with our low budget; we really need to heal what is below us. We do it because it's our job."[19]

*For more information on **Nance Klehm** and her projects, writing and courses, check out her website at spontaneousvegetation.net/.*

CHAPTER 5

# Phytoremediation

PHYTOREMEDIATION IS AN EARTH REPAIR TECHNIQUE that works with the natural capabilities of plants to repair and regenerate toxic soils, groundwater and surface water. Plants can help bind, extract, transform and clean up many kinds of pollution including metals, pesticides, chlorinated solvents, polychlorinated biphenyls (PCBs), explosives, radionuclides and petroleum hydrocarbons. Plants clean up these forms of pollution as far down as their roots can grow. Plants also help prevent wind and rain from carrying pollution away from the site to other areas.

Phytoremediation works best at sites with low to medium amounts of pollution. It is a low-cost, community alternative to many conventional and corporate land remediation practices. Instead of using excavators and dump trucks to remove tons of toxic soil and fill a site with new clean soil mined from elsewhere, on-site plants remove contaminants from the soil and store it within their plant tissues. In some cases, the plants themselves are then removed as toxic waste, while in other cases plants break down the chemicals and transform them altogether.

It is important to realize that there are many ways by which plants heal and restore toxic and damaged land. Understanding these different processes allows us to realize the different gifts that plants have to offer as healing allies. As we deepen our ability to understand nature's own strategies for self-repair, we can better assist in those processes that have been formed by a time and intelligence far greater than our own.

There are six common phytoremediation processes that plants bring to their environment.

## Phytoextraction

Plants take up contaminants — mostly metals, metalloids and radio-nuclides — with their roots and accumulate them within their stems and leaves. These plants are called hyperaccumulators.

## Phytodegradation

Plants take up and break down contaminants through the release of enzymes and through metabolic processes such as photosynthetic oxidation and reduction. In this process, organic pollutants are degraded and incorporated into the plant or broken down in the soil.

## Phytovolatilization

Some plants take up volatile contaminants and release them into the atmosphere through transpiration. The contaminant is transformed or degraded within the plant into a less toxic state and then released into the air. Sometimes the contaminant is released as is and then degraded by the sun.

## Rhizodegradation

In some cases, microbes in the soil break down contaminants into a less toxic state. In other cases, these microbes can completely destroy the contaminant. The root zones of certain plants create an environment that assists in this process. Therefore rhizodegradation is also called plant-assisted degradation.

## Rhizofiltration

Some plant roots can filter contaminated water by adsorbing contaminants into their root and plant tissue. Similar to phytoextraction, the plants themselves may become contaminated and have to be disposed of as special waste.

## Phytostabilization

Some plants can sequester or immobilize contaminants by absorbing them into their roots and releasing a chemical that converts the contaminant into a less toxic state. This method limits the movement of contaminants through erosion, leaching, wind or soil dispersal. It is often referred to as a "green cap."[1]

## Getting Started

For impacted communities and grassroots remediators, phytoremediation holds promise as an accessible and powerful form of earth repair. It has been used and tested by government agencies and universities worldwide, as well as by some environmental remediation companies.

While excavating toxic earth and burying it in a landfill can be done pretty quickly, phytoremediation can take anywhere from three to five years minimum for full cleanup to occur on a lightly to moderately polluted site. Further, the more different types of contaminants are present, the more complex and the longer your remediation strategy will take.

A recent field guide to phytoremediation created by the New York–based group youarethecity stated that the costs to phytoremediate a 2,500 square foot lead-contaminated site using brown mustard would be in the range of US$5,000–$8,000. Conventional excavation and filling of the same site would cost US $50,000–$100,000.[2] So it seems that phytoremediation can reduce the cost of recovering a toxic site to about 10 percent of the cost of industrial methods.

So how do we engage botanical interventions that repair and enliven depleted and damaged urban and rural places, transforming them into fertile and healthy common spaces and wild lands once more? The following steps are an introduction to using plants as powerful allies to extract, transform, stabilize and restore the land you are tending. The steps below and the bulk of this chapter refer to phytoremediating contaminated soil, not water.

## Step #1: Site and Capacity Assessment

In order to engage in effective grassroots phytoremediation work, you need to know as much as you can about the contamination on-site, the nature of the site itself, the resources available to you and your team, and the capacity of your group to maintain the site over several years.

Here are some questions to consider before digging in:

- What is the pollution on-site? Where is it? How deep?
- Where is the water table? What are the drainage and runoff patterns on-site?

- What needs to happen so that the site can support life?
- Is a fence or netting needed to keep animals and birds out so that they don't eat contaminated plants?
- What is the site use history?
- If you are working with plants to phytoextract contaminants, how will you dispose of the waste plant materials?
- What sort of native plants can you use on your site for remediation?
- What plants, if any, are currently on your site?
- What is your climate/growing season? This influences the speed of your remediation process.
- What grows well in your climate (e.g. mustard in cool climates, sunflowers in hot climates)?
- What is the pH of the soil?
- What is your remediation time line? How long can you sustain your effort? Are there deadlines or competing uses for the site besides remediation?
- Do you have the ability to constantly maintain (care for plants, volunteers, irrigation) and monitor site for three or more years?
- Will you use a combination of different remediation methods for your site?

These are important questions to answer. Some answers may not be available to you due to a lack of records or lack of funds for some of the more expensive tests. If that is the case, do your best. Be conservative in your estimates, creative with your approach and proceed.

### What Plants Can Tell You About Soil

If there is no vegetation on your site, that can be a strong indicator of contamination, compaction, soil pH being too high or too low and little to no soil biology. If your site is full of nettles, chickweed, groundsel and fat hen, that can be a sign of good soil fertility. If your soil is lacking nitrogen, there may be a lot of nitrogen-fixing legumes present, such as clovers and vetches. Some plants like bracken fern, plantains, sorrel, knapweeds, rhododendron and cranberries favor acidic soil conditions. Plants like perennial sow thistle, bladder campion, henbane and mustard can indicate more alkaline soil conditions. The presence of species like mosses, creeping buttercup and horsetail, sedges, rushes, marsh marigold, marsh orchid or flag iris can indicate waterlogged or poorly drained soil. Weeds can also signal that you are dealing with soil compaction.

Some plants indicate the presence of certain metals in the soils. For example, alpine pennycress (*Thlaspi caerulescens*) can indicate soils rich in zinc or cadmium. If hyperaccumulating plants, known for their high tolerance and ability to extract certain contaminants from the environment, are the only plants on your site, that may mean that something sinister is lurking in the soil. Some examples of these plants can be found in Figure 1 of this chapter.

### Getting the Soil Tested

There are several ways to find out what contaminants are on-site. If you have some money, you can get the soil tested to find out if there are heavy metals, hydrocarbons and chemicals present. Before you collect soil samples from your site to send off for testing, do some research on what possible contaminants might be lurking on your site. Soil tests are expensive, and laboratories generally test for requested contaminants only. If you have a general idea of what could be there, you will save money.

If you can find out what type of development or uses the site had, you can extrapolate, with a little research, what pollutants may be associated with such activities. Historical information can be found either by digging through your city or town archives (or state, provincial or federal bodies depending on what type of development it could have been) or by interviewing neighbors in the area, especially older folks who can take you a little farther back. Another helpful resource, if you live in the USA, is the US Department of Agriculture online Web Soil Survey.[3]

A *basic soil test* usually tests for pH and nutrient levels such as phosphorous, nitrogen and potassium. Most *soil contaminant tests* will test for organic compounds and fuel components like benzene, toluene, xylenes and other hydrocarbons, while *heavy metals testing* will test for lead, chromium, cadmium, nickel and zinc. *Toxics testing* will search for PCBs, mercury, cyanide, molybdenum, arsenic and selenium. Ask the laboratory, extension office or university you are sending your samples to what test makes the most sense for you and be as clear as you can about what you are looking for. The costs to test a soil sample for chemicals and metals can start around US$80–$100 depending on where you go, and can rise from there depending on what you are looking for.

Also, to get the best picture of where contamination is, you will want to gather several soil samples, taking them from at least four different areas for every 400 square feet of space. Some folks also recommend taking samples in a zigzag pattern across a site for better coverage. Samples should come from around six inches below the surface and should not contain any gravel, grass or foreign particulate matter. If you need to minimize costs, you can take soil samples from all around your site and combine them to create one composite sample. This would allow you to see if there was contamination on the larger site, but would not provide any specific locations of that contamination.

Test results will generally be mailed to you within one to two weeks. Ask for help analyzing the results from either the lab or someone at a local university so that you can actually make use of the information provided. Soil tests do not come with explanations, just numbers on a page.

If finances are tight, you may find it more valuable and fun to connect with your local university or college environmental sciences program to get free or cheap soil testing done, as well as water testing if also needed. Partnering is also a great way to engage more allies and people in your project, as well as to connect students with a socially responsible way of using their knowledge, skills and privilege. You can also reach out and cultivate a relationship with a local environmental consulting company or remediation business. These companies often have access to laboratories and testing equipment, as well as significant expertise on the pollutants themselves. Though many of these companies are driven by a strong profit motive and may not necessarily share your vision of ecological healing, you could be surprised by the resources and allies, formally or informally, that come your way from the more conventional sector.

### Gathering Your Materials and Skills

Phytoremediation is accessible at a community level because the tools required are cheap, available and can be solicited as donations or made from scratch. Some of the main materials you may want to have for your project are:

- shovels
- rakes
- wheelbarrows
- digging forks
- compost bins
- pots and seed trays
- irrigation (can be simple or complex — watering can, hoses, soaker hoses, sprinklers, drip tape)
- potting mix
- seeds
- plant starts
- compost and compost materials
- mulch
- greenhouse
- cloches or cold frames
- Reemay®/floating row cover (for colder climates)
- gardening gloves
- coveralls, respirator or face mask (depending on type of contamination you are working with)

When it comes to gathering compost and mulch materials, just because something is free does not mean it is the best addition for your site. The leaves of certain trees like black walnut can suppress the growth of plants, and using uncomposted horse manure can introduce exotic plant species onto your site that may later pose a weed management problem. Different compost and mulch materials can introduce other forms of contamination onto your site. Know the source of your materials (be it topsoil, leaves, grass clippings, wood chips or manure) and make sure that they do not contain antibiotics, pesticides, herbicides or other contaminants.

Municipal compost can be a free resource, but such compost may not possess strong microorganism populations, so you will want to inoculate it using compost tea and other practices. *Municipal biosolids* may also be a free or cheap resource offered to your project, and there are many conventional bioremediation projects that are using it. However, I suggest that you make your own compost. Municipal biosolids can potentially carry contaminants such as pharmaceuticals and heavy metals.

The skills needed for successful phytoremediation are knowledge and experience of growing and caring for plants (including the propagation of native and rare varieties) as well as how to build compost and soil. Additional expertise in the realms of soil science, hydrology and biochemistry all heighten your ability to diagnose the situation you are dealing with, but without these you can still proceed and get a lot accomplished. When it comes to this work, money is helpful to have, but our strength has been and will always be our social connections. Take time to consider all the potential allies and mentors in your community — professors and grad students on campus, local garden and native plant restoration clubs, coffee shops and breweries (who can offer you free compost materials), farmers (who can help with manure and plant starts), nurseries and garden centers (who can donate free tools and plant starts), seniors (who can tell you lots about the history of the site) — the list goes on! When doing the work of facilitating ecosystem regeneration, it doesn't hurt to have a dynamic and diverse social web around you.

## Step #2: Pick the Right Plants to Do the Job

After assessing your site and gathering as much information as you can, it is time to create a remediation strategy and determine which plants can repair your site. If you are looking to enlist microorganisms to bind heavy metals or break down chemical contaminants on-site, you may want to look at plants that have deep or extensive root systems. If you are looking to extract heavy metals in the soil, some plants are known to hyperaccumulate specific metals at a far higher rate than others. It is important to understand that hyperaccumulation of the metal by plants isolates and binds the problem, but does not solve it. Unlike organic chemicals that can be broken down to less harmful constituents, metals cannot be broken down; they can just be removed or sequestered in the plants.

What makes a good phytoremediating plant? One that is hardy, easy and quick to grow, produces a large amount of biomass, has a high evapotranspiration rate and a deep or extensive root system. A plant with these characteristics pulls up, stores and transforms (with organic chemicals) more contamination. Plants with deep roots are able to reach down to deeper contamination, and plants with an extensive root system are good at stabilizing contamination, as well as providing the ideal conditions for rhizodegradation. It is also critical that the plant is hypertolerant to the specific contaminant. Some plants are more tolerant and able to survive higher levels of metals, salt or chemicals. For example, alpine pennycress can accumulate up to 30,000 ppm zinc and 1,500 ppm of cadmium. These levels would poison most other plants, but alpine pennycress shows little to no toxicity symptoms at this level.[4]

### Phytoremediation Plant List

When reading Figure 1 below, be aware that most of the plants listed next to the heavy metals engage in phytoextraction, while those plants listed next to chemical contaminants engage in other processes, including phytoextraction, rhizodegradation, phytodegradation, phytostabilization and phytovolatilization. Also, the plants listed are ones that have been used or experimented with to varying degrees of success. They are not the only ones that can be used, and there are likely many plants not listed that could be useful.

You may want to start out with plants that are known to be tolerant of extreme site conditions (e.g. low or high pH, high concentration of particular contaminant, salt, low nutrient availability). Allow them to remediate the site while you also amend the soil (with compost, aerated compost teas and other soil amendments) in order to allow you to use a greater diversity of plant species later on.

Watch out for *salt*, as some contaminated sites have really high salt levels, and this can create an incredibly hard environment for plants. There are however some plants (such as tall wheatgrass, spear saltbush, lambsquarters and shepherd's purse) that are not only salt tolerant, but also good at removing salt from the soil.

Some folks feel that, even though they may at times be harder to propagate, *native plants* are at least equal if not better when it comes to phytoremediation, as they are more preferred by native microorganisms and have better-developed mycorrhizal relations. In many cases,

| Contaminant | Plants | Notes |
|---|---|---|
| lead | alpine pennycress (*Thlaspi caerulescens*)<br>sunflower (*Helianthus annuus*)<br>brown mustard (*Brassica juncea*)<br>geranium (*Pelagrium* spp)<br>corn (*Zea mays*)<br>tomato (*Solanum lycopersicum*)<br>buckwheat (*Erigonium* spp)<br>willow (*Salix* spp)<br>alfalfa (*Medicago sativa*)<br>wheat (*Triticum aestivum*)<br>water hyacinth (*Eichhornia crassipes*)<br>sheep fescue (*Festuca ovina*)<br>honey locust (*Gleditsia triacanthos*)<br>bald cypress (*Taxodium distichum*)<br>vetivergrass (*Vetiveria zizanioides*)<br>alyssum (*Alyssum wulfenianum*)<br>quaking aspen (*Populus tremula*) | Though there are many potential phytoextraction plants, some studies have found that lead is more easily phytoremediated by (adding specific amendments and a green cap to the soil to bind the lead) instead of phytoextraction.<br><br>Warning:<br>Water hyacinth is known to be an invasive species. |
| arsenic | Chinese brake or ladder brake ferns (*Pteris vittata*)<br>bentgrass (*Agrostis castellana*)<br>marsh fern (*Thelypteris palustris*)<br>white lupine (*Lupinus albus*)<br>poplar (*Populus* spp)<br>Douglas fir (*Pseudotsuga menziesii*)<br>rice (*Oryza sativa*) |  |
| mercury | sunflower<br>poplar<br>cottonwood (*Populus* spp)<br>rapeseed (*Brassica napus*)<br>water lettuce (*Pistia stratiotes*)<br>hydrilla<br>water hyacinth<br>willow | Warning:<br>Water hyacinth is known to be an invasive species. |
| zinc | alpine pennycress<br>tomato<br>bladder campion (*Silene vulgaris*)<br>sunflower<br>brown mustard<br>common duckweed (*Lemna minor*)<br>red clover (*Trifolium pratense*) |  |

Fig. 5.1: *Phytoremediation Plant Table.* CREDIT: LEILA DARWISH[5]

| Contaminant | Plants | Notes |
|---|---|---|
| zinc (cont.) | reed canary grass<br>  (*Phalaris arundinacea*)<br>wheat<br>water hyacinth<br>willow | Warning:<br>Water hyacinth is known to be an invasive species. |
| nickel | alyssum (*Alyssum lesbiacum*)<br>yellowtuft (*Alyssum murale*)<br>sunflower<br>mustards<br>milkwort (*Glaux* L.)<br>alpine pennycress<br>reed canary grass | |
| cadmium | mustard<br>cabbage (*Brassica oleracea*)<br>willow<br>field chickweed (*Cerastium arvense*)<br>miner's lettuce (*Claytonia perfoliata*)<br>wild chives (*Allium schoenoprasum*)<br>common duckweed<br>hydrilla<br>water hyacinth<br>sunflower<br>tomato<br>tape and eel grass (*Vallisneria* spp)<br>alfalfa<br>alpine pennycress<br>common yarrow (*Achillea millefolium*)<br>purple foxglove (*Digitalis purpurea*)<br>northern starwort (*Stellaria calycantha*)<br>holly (*Ilex* spp) | Warning:<br>Water hyacinth is known to be an invasive species. |
| chromium | brown mustard<br>goat willow (*Salix caprea*)<br>basket willow (*Salix viminalis*)<br>birch (*Betula* spp)<br>smooth water hyssop<br>  (*Bacopa monnieri*)<br>water hyacinth<br>sunflower<br>tape grass<br>alpine pennycress<br>creosote bush (*Larrea tridentata*)<br>lodgepole pine (*Pinus contorta*)<br>gray alder (*Alnus incana*)<br>red alder (*Alnus rubra*) | Warning:<br>Water hyacinth is known to be an invasive species. |

| Contaminant | Plants | Notes |
|---|---|---|
| chromium (cont.) | European alder (*Alnus glutinosa*)<br>black cottonwood (*Populus balsamifera* L. ssp. *trichocarpa*) | |
| copper | brown mustard<br>water hyacinth<br>common duckweed<br>smooth water hyssop<br>bladder campion<br>sunflower<br>alpine pennycress<br>creosote bush<br>willow | Warning:<br>Water hyacinth is known to be an invasive species. |
| selenium | two-grooved milk-vetch (*Astralagus bisulcatus*)<br>rapeseed<br>barley (*Hordeum vulgare*)<br>desert princesplume (*Stanleya pinnata*)<br>willow<br>fly agaric (*Amanita muscaria*) | Some plants phytovolatize selenium and some extract it. |
| molybdenum | alpine pennycress | |
| cobalt | alpine pennycress | |
| aluminum | barley<br>fava bean (*Vicia faba* var. *equina* Pers.)<br>hairy goldenrod (*Solidago hispida*) | |
| silver | brown mustard<br>rapeseed<br>willow | |
| cesium | red maple (*Acer rubrum*)<br>larch (*Larix* Mill.)<br>spruce<br>redroot amaranth (*Amaranthus retroflexus* L.)<br>amaranth (*Amaranthus* spp)<br>tulip tree (*Liriodendron* L.)<br>monterey pine (*Pinus radiata*)<br>ponderosa pine (*Pinus ponderosa*)<br>coconut palm (*Cocos nucifera*)<br>cabbage<br>beans (*Phaseolus* spp)<br>spiderwort (*Tradescantia* L.)<br>reed canary grass<br>brown mustard<br>beet (*Beta* spp) | |

| Contaminant | Plants | Notes |
|---|---|---|
| cesium (cont.) | quinoa (*Chenopodium quinoa* Willd) corn sunflower Russian thistle (*Salsola* L.) field chickweed tall fescue (*Schedonorus arundinaceus*) perennial ryegrass (*Lolium perenne*) white clover (*Trifolium repens*) | |
| plutonium | red maple tulip tree | |
| uranium | sunflower cabbage, redroot amaranth oak (*Quercus* spp) corn black spruce (*Picea mariana*) juniper (*Juniperus* spp) bladder campion | |
| strontium | monterey pine ponderosa pine forest redgum (*Eucalyptus tereticornis* Sm.) sunflower water hyacinth | Warning: Water hyacinth is known to be an invasive species |
| hydrocarbons (PAHs — polycyclic aromatic hydrocarbons and TPHs — total petroleum hydrocarbons) | hybrid poplar (*Populus deltoides x Populus nigra*) pine (*Pinus* spp) tall fescue red fescue (*Festuca rubra*) black locust (*Robinia pseudoacacia*) willow sunflower white clover yellow sweet clover (*Melilotus officinalis*) red clover Bermuda grass (*Cynodon dactylon*) western wheatgrass (*Pascopyrum smithii*) blue grama grass (*Bouteloua gracilis*) buffalograss, (Bouteloua *dactyloides*) Canada wildrye (*Elymus canadensis*) perennial ryegrass California white sage (*Sativa apiana*) switchgrass (*Panicum virgatum*) red mulberry (*Morus rubra*) alfalfa | Hydrocarbons with shorter carbon chains are easier to phytoremediate than more complex hydrocarbons. If you are trying to pull it up and out of the soil, some studies have found that this can be done more effectively in soil with low organic matter. The different plants listed here interact and work with different hydrocarbons, |

| Contaminant | Plants | Notes |
|---|---|---|
| hydrocarbons (cont.) | apple (*Malus domestica*)<br>cat claw (*Mimosa monancistra*)<br>pumpkin (*Cucurbita pepo*)<br>rice<br>mushrooms<br>microorganisms | as well as in different ways to remediate these contaminants. |
| volatile organic compounds (e.g. benzene, toluene, ethylene, xylene) | poplar<br>white rot fungus<br>    (*Phanerochaete chrysosporium*) | |
| TCE (trichloroethylene) | pine<br>willow<br>alfalfa<br>bermuda grass<br>hairy goldenrod<br>red mulberry<br>hybrid poplar | |
| naphthalene | white clover<br>tall fescue<br>ryegrass<br>red clover | |
| MTBE (methyl tertiary butyl ether) | alfalfa<br>white rot fungus<br>hybrid poplar | |
| PCBs | pumpkin<br>white clover<br>white rot fungus<br>rose (*Rosa* spp)<br>red mulberry<br>garden orache (*Atriplex hortensis*)<br>osage orange, (*Maclura pomifera*) | |
| DDT | zucchini<br>pumpkin and squash varieties<br>    (*Cucurbita* spp)<br>common duckweed<br>hybrid poplar | |
| perchlorates | sweetgum (*Liquidambar styraciflua*)<br>loblolly pine (*Pinus taeda*)<br>willow | |
| salt | tall wheatgrass (*Thinopyrum ponticum*)<br>common orache (*Atriplex patula*)<br>shepherd's purse (*Capsella bursa-pastoris*) | |

| Contaminant | Plants | Notes |
|---|---|---|
| Salt (cont.) | lambsquarters/fat hen (*Chenopodium album*) Jerusalem artichoke (*Helianthus tuberosus*) | |

native plant species are better adapted to live in your climate. You may want to experiment with this a bit. If you are not using native plants, be sure to consider the impact on your local environment of using potentially *invasive plants*. Also, be wary of *transgenic* and *genetically modified plants*. Universities, research centers and land reclamation companies are breeding specialized varieties of phytoremediator plants with amplified extraction abilities. If these plants are available to you, consider the impacts they might have on the environment, and try to find out more before using them on your site.

Some of the plants that are good remediators may be available to you, but others may be harder to find. In this case, you may want to connect with specialty nurseries in your area or order seeds online. Universities may also be good allies for this. Your local native plant society can also prove really handy. Consider working with other folks in the community to start a local operation for growing key plants to assist in phytoremediation efforts. Hopefully, as more communities

Fig. 5.2: *Sunflowers — an easy-to-grow remediator.* CREDIT: KAJA KÜHL

start engaging in phytoremediation and experimenting with plants, we can share more local solutions that will likely prove to be more effective than a one-size-fits-all approach.

### Tree Remediation

Figure 1 above lists several trees that have been used for remediation. Some folks refer to using trees in this way as *tree-mediation*. Fast growing trees with longer roots (such as poplars, willows or black locust) go farther than plants can, accessing deeper contamination. Poplars, for example, grow quickly, are hardy and make the environment more hospitable for other plant and tree species. Their roots are able to absorb as much as 30 gallons of water per day, which means they can pump up a lot of contaminated groundwater to be phytoremediated.[6] If the trees are working with heavy metals that they hyperaccumulate, then these heavy metals may become bound into the trees' long-lived wooden structure. If toxins are present at a low enough concentration, we may be able to safely use the wood for building or craft purposes, reducing the need for hazardous waste disposal. It would be great to test your trees, to see where they are accumulating contaminants. Determine if it is safe to use their leaves for mulch and their wood for chips or other materials.

Trees are healing investments. They can take a while to get to that place where they have long enough roots to reach the deeper contamination and sufficient biomass to really rock out the rhizodegradation and phytovolatilization of certain contaminants. Once in place and given the time to become the strong healers that they are, trees are powerful at their remediating work, allowing us to access contamination that would be difficult to reach and moderating the site to become more livable and protected for other plants. Consider what you will do in the initial years before the full remediative force of your trees has yet to come online. Consider engaging in microbial remediation (at least for increasing fertility of the site, and with some contamination breakdown depending though on how deep that contamination is). You could also sheet mulch, or simply mulch the site around your trees to keep soil from moving too much and for water retention. You could build some raised beds to safely grow some veggies while your trees get established. Or you could grow cover crops to build fertility and organic matter on the site. Once your trees get established, though, you will be dealing with a more stable perennial remediation system, with its own maintenance needs.

## Step #3: Prepare Your Site for Growing Phytoremediators

Once you have your plants or seeds ready, you need to consider whether they can live on the site you are planting them into.

### It All Starts With Healthy Soil

Is the soil compacted? Does it look rough and barren? If so, you may need to do some site prep to make it easier for your plants to thrive. To deal with compaction, use some rebar or a digging fork (depending on how deep you need to go) to poke holes throughout the site to help break it up and allow for more aerobic soil conditions. You can also drench a newly aerated site with compost tea and bring in some compost and other soil amendments (to deal with nutrient deficiencies or modify the soil pH).

Using compost (see Chapter 4) and spraying some aerated compost tea will inoculate your site with beneficial microorganisms and bring back the powerful soil microorganisms that will not only help your plants grow better, but will also work to break down and bind certain harmful chemicals and hydrocarbons in the soil. Compost will also provide much needed organic matter, as well as provide a better growing medium and soil texture. Adding compost can also dilute the volume of the contaminant in the soil, as well as reduce the availability of the contaminant for uptake by plants. If you choose to go the route of phytoextraction (be it for metals or for chemicals like PAHs), be aware that adding organic matter (via compost) to your soil could make it harder to pull up the contaminants as the organic matter could have a binding effect.

### Adjusting Soil pH and Adding Amendments to Extract or Immobilize Heavy Metals

If you are working with heavy metal contamination, do you want to extract the contamination or do you want to immobilize it? Depending on which route you choose to go, you may have to amend your soil pH. We all understand what it means to extract the metal, but in this context what does immobilization mean? *Immobilization* is the process by which the solubility and the movement of heavy metals in soil is reduced and they are made less bioavailable to be accumulated by plants. Before going much further, you should select your remediation route and decide whether you will be extracting contaminants, immobilizing contaminants or using both strategies at different times for different contaminants.

Different metals react to different soil conditions and pH differently, depending on whether they are *cationic metals* or *anionic metals.* Soil must be acidic (have a pH lower than 7) for plants to easily remove/extract/accumulate lead and other cationic metals such as cadmium, copper, mercury, thallium, zinc, antimony and barium. If the soil is more alkaline, cationic metals will become less mobile and will less easily be taken up into plants. Anionic metals like arsenic and chromium become more bioavailable and easy to extract via plants when soil is alkaline/basic (has a pH higher than 7). If soil is more acidic, anionic metals will be less available and more bound.

You may find more than one metal contaminating your soil. Often, lead and arsenic are found together, as are zinc and cadmium. If you are in a situation where you have both cationic metal (lead) and anionic metal (arsenic) contamination in your soil, you may have to extract contaminants in several steps, rotating between acidic soil conditions and basic conditions to remove them both from your soil, and you will need to be aware that what immobilizes one kind of metal may make the other more mobile.

Determine what the pH of your soil is, either from looking at your original soil test results or doing a soil pH test yourself, which is cheap and easy and the equipment can be purchased from most garden centers or specialty nurseries. A pH meter can cost anywhere from US$30–100.

If you are trying to favor the extraction of lead (and other cationic metals), add materials to the soil that will lower its pH, such as organic sulfur, pine needles, peat moss, sawdust, leaf mold and coffee grounds. Doing this will also immobilize anionic metals. If you are trying to favor extraction of arsenic (or other anionic metal), then add materials to the soil that raise its pH, such as some form of organic phosphorous (e.g. bat guano or fish bones), chalk or ground limestone, wood ash, bone meal, crushed shells or marble chips. Another way to add phosphorous to the soil could be to plant a cover crop of buckwheat. This approach would take longer than simply amending the soil. To add phosphorous to the soils she works with, Nance Klehm incorporates weeds with high phosphorous content into her compost and then uses that compost, allowing the microbes to digest, share and circulate. She also uses plant extracts of these weeds. Some of these weedy heroes are: dock, chickweed, sorrel and vetch.[7]

Extremes in soil pH can also create conditions on your site that hamper the growth of your remediating plants. For example, aluminum is the third most abundant element in the Earth's crust, and it is a major component of clays in soil. At neutral or alkaline soil pH, aluminum does not pose a problem for plant health. But at lower pH and in more acidic soils, a form of solubilized aluminum can poison and damage plant roots. If you are dealing with a site with acidic soil conditions and also a high clay content, high levels of aluminum may create a toxic environment and hamper the growth of your plants. If this is the case, raise the pH of your soil via amendments and compost. Copper, and zinc and nickel can also produce toxic conditions for your plants and prevent them from growing when the soil is acidic, so amending the soil to raise the pH if these metals are present can really help your plants grow.

Amendments such as phosphorous or iron can also react with some metal contaminants (like lead, zinc, copper and cadmium) in the soil to create more stable, less soluble and non-toxic compounds. Please refer to the "Food Gardens and the Dreaded Lead" section later in this chapter for more information. Using amendments to change the nature of your contaminants can get a little chemistry heavy and be quite complicated, so proceed with caution and do more research.

### Irrigation and Watering

Put some thought into how you will irrigate your site so that your plants can grow. Is there a water source on-site or nearby that you have access to? Will you install irrigation or rain barrels? Use sprinklers, hoses or watering cans? If soil contamination is a big deal on-site, watering methods that cause a lot of spray and splash may not be the best way to go. You may want to choose soaker hoses, drip tape or other forms of more low-to-the-ground irrigation. If money is an issue and hoses and regular sprinklers are all you have, then that's fine. Mulch the soil around your plants to reduce splash back, and avoid powerful sprays. Watering cans would also work fine.

## Other Considerations

There are certain tough and hardy plants that thrive in very disturbed sites, and their job is to hold the soil, accumulate nutrients, fix nitrogen and create a more moderate climate in which more plant species can thrive. Many of these plants are common weeds and powerful healing herbs like dandelions, plantain, mullein, comfrey, yarrow, red clover and many types of grasses and legumes. Nitrogen-fixing trees, such as black locust and red alder, will grow quickly, help other plants and trees grow on your site and provide shade and shelter for more sensitive plants. You may want to consider working alongside these plants and allowing them to do their important work of stabilizing and making your site more suitable for the more specific remediators you will be using later on.

Take some time to consider how you will fence or protect your site. You want to make sure that animals and birds, who may be attracted to the plants growing on your site, don't eat hyperaccumulating plants, as you don't want to poison them and also because you need your plants to be living to do their work. You can harvest the edible parts of plants immediately (e.g. corn cobs or sunflower heads) to avoid birds, squirrels or children consuming them. A simple fence or netting should suffice, also serving the other function of letting people know that the site has risky and fragile elements.

## Step #4: Figure Out How To Dispose of Toxic Material and Plants

If plants extract metals from the soil, once they have completed a growing cycle you could be left with plant residue laden with these contaminants that can technically be considered toxic waste. You need to treat this material with caution. You should definitely get your plants tested when you are ready to harvest them, to see whether they are too toxic to compost or dispose of regularly. Some folks have found that plants are within acceptable levels for municipal landfilling; you never know unless you test. If you are unable to afford such testing, then I would treat the plant waste as toxic and dispose of with the appropriate precautions. You definitely cannot use questionable plant wastes for compost and soil creation on your site; you may spread pollution around instead of removing it.

Where will you be sending the waste plant materials? There is no perfect, feel-good answer for this one; there are just different pros and cons to weigh. Three main options are available.

### Municipal Landfill

If your plants have been tested and shown to be within acceptable limits, they can safely be disposed of as normal waste or even composted.

### Hazardous Waste Site or Incineration

If your plant waste tests toxic and above municipal landfill guidelines (or you are dealing with really toxic stuff that you would never dream of putting in a landfill), collect it (or have a company collect it) and then transport it to a *hazardous waste site*, a landfill with special type of liner that hopefully makes sure that no seepage can occur.

In some cases, there may also be the option to send plant waste to an incineration facility where the plants are burnt down and the metal recaptured. This is not necessarily a financially feasible option for community projects. Some experimental treatments have added chemicals that leach out accumulated metals so that they can be collected and used again. In some ways, this is promising work, but both incineration and leaching methods create their own environmental impacts.

All in all, if you send your plant waste away to a hazardous waste site, you will be effectively removing the metals from your site, though contributing to the creation of a sacrifice zone elsewhere. But on the upside, by using phytoremediation instead of the conventional method of landfilling all the contaminated soil, you could create up to 95 percent less toxic waste.

### Store and Concentrate Plant Waste On-Site

If there is an appropriate place on-site to store plant wastes safely, then consider concentrating them in that part of the site. If you choose to keep waste on-site, you need to understand that you are creating a sacrifice zone there. This is not necessarily the worst thing, considering that at the outset your whole site may have been a sacrifice zone. If you choose this method, please pick this containment/disposal area carefully. You should only use this method if your site has a location with appropriate characteristics:

- Dispose of your plant wastes at a location on your site that is chosen for its physically stability.
- If you are aware of where there is water on your site, do not put your waste depot near it.
- Choose a place that is protected from wind and out of easy access for animals, birds and is clearly marked and identified for people.
- Make sure your wastes are covered. For this you should dump your plant waste into a sturdy, animal-proof structure/bin with a lid.
- Consider purchasing a liner.

You can treat this location as a compost area of sorts, one that you keep covered, in one spot and do not spread on your vegetables or plants. By composting your wastes, you won't get rid of the contaminants, but you will concentrate them in smaller and smaller quantities of earth, which is good. In the end, if you've managed to transform a whole toxic site into a clean site with a small and well contained toxic waste storage zone, then you are making progress.

> In all three situations, be mindful with how you and your team handle waste. Take health precautions when you are harvesting plants, bagging the waste or turning the compost, especially for the first few rounds of phytoremediators. Wear gloves, bring a change of clothes and consider using a face mask.

## Step #5: Plant the Plants, Grow the Plants, Test the Soil/Plants and Start Again!

Finally, after all the initial setup work, comes planting and growing plants. During this phase of the project, you will have to do a lot of maintenance work and give a lot of care to the plants, especially in the beginning as they are getting established. If you have a few false starts where you lose plants or they just don't take, do not lose heart. Just go back to Step #3 and work to modify your soil (e.g. pH, organic matter) and microclimate (e.g. exposure, temperature, moisture). Also consider working with some different plants until you find a good fit. For some plants, direct sowing of the seed will yield the highest success. For others, they will take better if transplanted. Consider inoculating your seeds or plant starts with mycorrhizal fungi to increase their resilience (see Chapter 4). Consider using floating row cover/Reemay® or building a movable cloche to help your plants get started if your climate is prone to big temperature fluctuations.

Make sure to put up signs explaining to the public what you are doing on-site and to warn/advise people against eating the plants or allowing their animals to forage there. This is important to protect folks and animals from ingesting plants that are accumulating toxins (therefore making some of that contamination more available to the food chain than it was before). It is also helpful to keep the community informed, aware of potential risks and of the rad work you are doing, as well as to show the general public that you are concerned for their safety. Protective covering, as well as signs, also provide protection to your plants, which may need tender loving care in the beginning and can only do their work if they are alive (not being trampled, crushed or eaten).

If you are working with hyperaccumulator plants to extract metals, once they have completed a growing cycle, harvest them and dispose of them safely. There is some ambiguity around how long you should let them grow before they have taken in their maximum amount of toxins — some sources say after 14 weeks you should harvest and then replant. This will vary depending on the plant and your climate. If you can conduct another round of soil tests at this point, or even send off some of your plant wastes to be tested, you will see if what you are doing is having an effect and how far along you are. If you are working with more perennial plants to bind and break down contaminants, then test the soil around them after a growing season to see if progress has been made. Through all these crop rotations, pay attention to the plants and the soil. Do your best to provide excellent care. You may choose to stick with a few plants for your whole effort, or you may choose to experiment with different plants.

Then, if you live in a northern climate with a snowy winter, mulch or cover your soil. Wait until spring to amend the soil again with compost tea (you can never have too much beneficial soil bacteria and microbes), add some clean compost and plant another round of either the same plant(s) or some different ones. If you live in a northern climate with a rainy winter, amend your soil with compost tea and plant a cold-tolerant hyperaccumulator plant or cover crop that can handle low temperature. You may want to mulch as well. If you live in a warmer southern climate, then add some compost, compost tea and maybe a few other amendments and keep going full speed ahead with remediation. Phytoremediation takes more time in colder climates than warmer ones, so take that into consideration.

### Sustaining Community Interest and Volunteer Support

Many of us have been a part of community garden projects where everyone thinks it's a great idea, but when it comes to the hard work the ranks thin out considerably. Sustaining interest and effort is definitely a challenge in community projects, especially for bioremediation. Many folks approach these projects with unrealistic and glamorized expectations of the work involved and can lose interest when confronted with a more humble, repetitive, messy and sometimes physically demanding reality. Also, the time needed to bioremediate a site to the point that it is safe and healthy is at least two to three years, likely more. In the case of a phytoremediation project, folks have to take care of plants that they will not get to use. It helps to be internally motivated, community minded or be part of a team which can keep putting the work in context and the dream alive.

One of the most important pieces to pulling off a grassroots phytoremediation project is to make sure you have the sufficient labor to do so and sustaining the interest through the remediation process. There are two main suggestions I offer here.

### Hybridize Your Phytoremediation Site with Edibles in Raised Beds

Engage in a hybrid phyto operation on-site. This idea came from Kaja Kühl at youarethecity in New York. She is currently working on a phytoremediation project called "The Field Lab" on a piece of land on the community farm La Finca del Sur in the South Bronx. This particular piece of land has some heavy metal contamination, and she and others are testing the ability of four different plants to remove them: brown mustard, basket willow, sunflower and mugwort.[8]

Kühl suggests that you could divide your site in two pieces: put in raised beds with clean soil/compost on one side, and plant the other side in phytoremediating plants. Grow veggies in your raised beds, which will give folks more incentive to come and harvest, which puts them on the site so that they can also check on the phytoremediating plants. This is a clever idea, as both veggies and phytoremediation crops require similar infrastructure in terms of tools, greenhouse and irrigation, and folks can engage with the site in different ways. After two to three years, you can move the raised beds and then plant phytoremediators in their place and switch veggie production to the previously remediated half.

## Engage the Community through Events, Workshops, Trainings and Job Opportunities

The best volunteers for a grassroots bioremediation project are local folks who live near the site and have a vested interest in that land becoming healthy, active and whole again. Engaging and involving the community is key to the success of the site. If the community understands, encourages and is excited about the work that you are doing on a site, you will have more folks willing to share their knowledge and skills, as well as help out with watering, checking on plants, donating useful materials and money and generally keeping a watchful eye out for the land and any infrastructure on it. This is very important. If the majority of the community you are working in does not understand or support what you are doing, it is disrespectful to push past community concerns, and it is very important to engage in education, outreach and dialogue to figure out what is the best way forward and how to address these concerns and build support for the project.

Community engagement also means that if you come into conflict with the government or company and they try to unjustly stop your project, then more folks will join you and mobilize to keep the project going and the site activated. Secondly, maintaining interest in the project and putting your work into the larger context is key to recruiting, motivating and retaining folks to help. Creating a dynamic culture around the work by involving skill exchanges and learning opportunities keeps people interested and inspired. Celebratory and fun events help too. Also, having posters, giving presentations and sending out the occasional email sharing different roles and work/volunteer opportunities is key in catching new interested folks.

## Step #6: Transitioning from Phytoremediation to Food Cultivation, Parks, Gathering Spaces or Re-Wilding

Some sources say that you should plan on doing two to three years of phytoremediation work on a site with low to medium contamination. Others say up to 30 years, depending on the type and magnitude of the contamination and the climate you are working in. The only way to know how well the healing is coming along is to test your progress.

One of the goals of our work is to facilitate and support ecological processes in the living systems you are working alongside to restart and hopefully reach a critical point where they can continue without your intervention. If you are attempting to re-wild the site, plant native grasses, plants and trees and let nature take care of the rest. Grasses (not Astroturf lawn) are great as their comprehensive root system binds contaminants; they also provide excellent habitat for the many beneficial microbes that will continue breaking down and binding any remaining contaminants. Trees will moderate the environment, create their own mulch and can store a lot of the remaining contamination in their woody fibers, which take a while to break down. Figure out the different stages of succession on your site and try to plant the appropriate plants for the stage that you are in. You can also continue

## Let Birds Re-Wild and Revegetate Your Site

This idea comes from one of my permaculture teachers, Oliver Kelhammer. Put up birdhouses, bird feeders and if you can a little birdbath (source of water) to attract these amazing beings to your site. Why? Because you want them to poop all over your site. Confused as to why that's a great thing? The bird poop contains all the lovely seeds they have eaten in the forests and meadows around your site and brings these species to you — with minimal effort on your part. Not only will birds start off re-wilding and revegetating your site with a whole host of native and non-native species, their poop is rich in nitrogen and phosphorous which will help fertilize your site.

bioremediating the site by applying aerated compost tea and compost frequently.

If you plan on food cultivation, then continuing to generously apply compost and compost teas on a regular basis will not only help with the fertility of your site, it will also continue to heal the land. At this point, if you haven't already done so, consider building up soil on the site. You can do this by laying down sheet mulch or installing raised beds. Basically you have pulled up as many toxins as you can, so now you want to bind the rest or make them less accessible to plants, animals and humans. Also, if your site is located in an area where there may be a lot of air pollution (like near a busy road or industrial area), consider laying floating row cover or a movable cloche over your food crops to avoid deposition of contaminants onto them.

## The Freeway Food Forest: Hayes Valley Farm

A freeway-ramp-turned-food-forest in the heart of San Francisco, the Hayes Valley Farm is a 2.2-acre urban farm and education center built on a former central freeway ramp in San Francisco. In 1989, parts of the central freeway collapsed as a result of the Lomo Prieta earthquake. The lot remained vacant and closed up for 21 years. Finally in 2009, the mayor of San Francisco signed a new urban agriculture directive, and the City finally authorized the site for interim use. The farm opened in 2010, and 150 people came out to volunteer at the first work party. Today, the farm is completely volunteer-run and donates a lot of its produce to charity and to its volunteers. I had the opportunity to interview Jay Rosenberg, co-director/co-founder of Hayes Valley Farm, who is urban farmer, educator and permaculturalist. He is also part of the organization "49 Farms" whose goal is to get 49 farms set up, one in every square mile of San Francisco.

"I took time to learn about bioremediation, habitat restoration, ecology and the systems that bring back life. In those systems, you get a phenomenal growth of food as a byproduct," said Rosenberg. Soil on the lot was quite toxic from over 50 years of brake dust, lead-based oils and carbon monoxide emissions from cars. The farm team did soil testing of the site. They found the parts on the site that were the most contaminated and left them. They built up the rest of the site with six to ten feet of organic matter (such as horse manure, wood chips, cardboard), which then composted down to be three to four feet thick. "In those places, soil samples came back healthy and delicious organic soil, and that's where we grow food. Compost, compost tea, mulch. The solution is so simple: add organic matter. The big OM! That's where it starts, and cities produce a lot of waste," stated Rosenberg.

Instead of removing the concrete to get at the contaminated soil below, the farm team took a different approach. "We used the existing concrete and pavement to seal most of the toxins in. One of the first things people thought we were going to do was remove the pavement. There is something romantic about breaking up pavement and revitalizing the earth," explained Rosenberg. "But what I've learned here is that pavement can be a phenomenal seal. By layering horse manure, cardboard and wood chips right on top and by laying it on really thick, we are building ecosystems up, and that is more romantic than cutting through the concrete and then having to spend millions of dollars in hazardous waste removal of whatever else is under there after we open it up."

Fig. 5.3: *Before shot. Looking down the highway off ramp at the site before sheet mulching.* CREDIT: JAY ROSENBERG

Besides a massive sheet mulching effort, berms were formed and planted down the hillside with fava beans to help cover the new earth, build the soil and yield a tasty crop. "We then planted sunflowers and beets. Then we planted an ecosystem of trees. The site started healing itself," said Rosenberg. "When we talk about toxic spaces, it's actually a combination of remediation and restoration that has to happen. To actually do the work of cleaning these spaces, it's about restoring the ecosystems. That's where the mushrooms, the soil, the worms, the birds and the bees come in. They all come together and they do the work we can't do, and they do it full time. They create ecosystems, and they heal these places in ways we haven't been able to grasp with our military industrial solutions."[9] The site has since become a fertile gathering place for nourishment and skill building, with everything from potted fruit trees, to mobile cob ovens, a greenhouse and bee hives.

Sadly, the farm's fertile days on the site may be coming to an end soon, as the current landlords have plans to start building some condos on the site. However, Rosenberg seemed quite optimistic and resilient about the whole thing. It was always known that the farm would be a temporary fixture, a placeholder for a condo development, which was one of the reasons why it was given permission to set up in the

Fig. 5.4: *Laying down cardboard and getting ready to sheet mulch and build soil at Hayes Valley Farms.* CREDIT: JAY ROSENBERG

Fig. 5.5: *Hayes Valley Farm today.* Credit: Jessie Raeder

first place. Everything was designed with mobility in mind, for the day when the farmers would have to pack up and move to another site. The fruit trees were planted in pots, the cob ovens built on pallets, and the greenhouse is easy to dismantle. Even the new soil could be hauled away to a new site if necessary.

Though some folks might think it an exercise in futility to go to all the hard work of creating a fertile and lush farm only to have it eaten up by concrete and condos, the fact remains that in the three years the Hayes Valley Farm has been operating, they have produced tons of local and healthy food for the community as well as provided habitat for many living things to thrive and jive. They've also shown that a freeway can become a food forest, and that is a powerful dream to enact in a city. The idea of the temporary and mobile farm that hops from vacant lot to vacant lot is one that embraces some of the tough realities and constraints of the urban jungle. In many ways, our cities still feel that condos and malls are more important than the gathering places, farms and gardens that feed our communities. Until that reality changes and our cultures shifts, farms like Hayes Valley will bravely bounce around, finding ways to facilitate fertile convergences in whatever abandoned or vacant lots they find.

*For more information and updates about* **Hayes Valley Farms,** *check out hayesvalleyfarm.com. For more information about 49 Farms, check out 49farms.org.*

---

## Sheet Mulch 101

*Sheet mulching*, also known as lasagna gardening, is a popular permaculture technique for both building soil volume and fertility with minimal disturbance to the original soil structure. It allows you to build layers of clean soil atop possibly contaminated soil in a way that still allows for biological processes to happen in the lower layer of soil. Whatever you plant into your sheet mulch (as long as your sheet mulch is deep enough and your plant doesn't have a long taproot) will grow and be quite safe to eat.

You can begin any number of ways, but the trick is to essentially lay down cardboard and then alternate with nitrogen layers and carbon layers to create active compost. The purpose of the cardboard is to block out light to the grass or weeds below, as well as to act as a carbon layer. Build up your layers, thick and deep. After a few of these different layers, you cap off the sheet mulch (at least until the next season) with either leaves or straw.

You can either wait three to six months for it to decompose into rich compost and create delicious and clean soil for your plants to be planted into, or you can plant into it immediately. If you need to plant into it immediately, you have three options:

- For your finishing layer of the sheet mulch, spread a thick layer of finished compost or clean topsoil.
- Push away the mulch from the hole you will be planting into, fill each hole with finished compost and then plant into that.
- Certain crops can handle unfinished compost better than others. Heavy feeders like squash, beans, corn and tomatoes do quite well in fresh sheet mulch.

In conventional remediation and some urban farming cases, another method to grow food on questionable soil is to put down a barrier or liner made from some synthetic fiber and then truck in clean soil to put on top or into raised beds. Or to build raised beds with clean soil or compost in them; this method can work well and is popular, especially when folks are unsure of the contamination levels below. But it can be expensive, and why go to the trouble of trucking in soil when you can build your own?

Besides food production or re-wilding, there are a few other options for remediation sites. If you are still worried that you have not been able to remove all the contamination or that it may not be suitable for food crops, consider using the site for other activities that will increase its fertility and provide habitat for different organisms. For example, if you have a rather wet site, you could plant it with lots of willow, and use that willow to weave baskets, fences and furniture and to create rooting hormone. You could plant other trees like black locust, which is great for construction as it is rot resistant, or bamboo (depending on where you are in terms of cautions around invasiveness) as it is also great to use for building. You could plant your site as a *nectary*, full of bee forage and nectar flowers (although I wouldn't do this on a site with moderate to high contamination). You could also keep part of the site as a *community composting operation*, which would build soil on-site, utilize local resources, divert local organic waste streams and provide your site with tons of beneficial microorganisms to continue bioremediation. You could throw in some inoculated mushrooms logs into the mix to grow mushrooms on clean materials. There are a lot of options for how your site can be transformed into a fertile place that contributes to both human well-being and planetary wellness.

Finally, consider how to invite people into your site. Maybe you want to get folks together to build a community noticeboard or some neat community goods exchanges (like a poetry station, free tea station or community book exchange). What about a beautiful cob bench or two for gathering, sitting and finding peace in the space? Or a community cob oven for pizza parties and bread making? Or different art installations? Or a community nursery? Maybe a play area for children (if the site has been significantly remediated to safe levels)? What about signs explaining the site, its history and its healing process?

What about a community mural or mosaic? The sky is the limit, and use the creation of different features as learning opportunities for anyone interested through community work parties and workshops. Seek out skilled members of your community to lead events in ways that enrich the intergenerational fabric and connectivity of the project. At the end of the day, you want to create a regenerative landscape, one that enhances the relationships and connectivity between people and place, and people with each other.

## Radical Urban Gardeners of The Purple Thistle Centre — Planting a Food Forest in East Vancouver

BY ADAM HUGGINS

The Purple Thistle is a youth-collective-run space for arts and activism, offering free art supplies, studio space, resources, classes, workshops and mentoring opportunities for youth in East Vancouver, BC, Canada. The Youth Urban Agriculture Project reclaims trashed marginal land in the industrial lowlands near Clark Drive and turns them into beautiful, bountiful permaculture gardens for growing

Fig. 5.6: *Purple Thistle Food Forest sign.* CREDIT: ADAM HUGGINS

food, medicine, materials and community. Growing from a handful of idealistic guerilla gardeners into a coordinated farming collective maintaining two urban gardens, one urban food forest, a wetland remediation project, beehives and an apothecary of herbal medicines, we value creativity and diversity above all, maintaining our gardens collectively (sharing the work and the harvest) and welcoming any and all who wish to get dirty with us!

As we were plotting and designing our community food forest project, we learned that the land we are growing on used to be the marshy banks of False Creek before railway companies filled them in during World War I. Bordering a disused train yard, our site is an excellent example of the terrain created by decades of industrial activity: soils that consist almost exclusively of sand and gravel, mountains of garbage and toxic waste and resilient native ecologies struggling to burst through the cracks — in our case, a battered wetland ecosystem (a striking example of what the marshland may have been before) hugging the edge of a proposed food forest. Thus far, we have removed hundreds of pounds of trash from the wetland and begun to establish native soil builders and remediators, with a view to creating a system capable of supporting a diversity of life, filtering the toxic water trickling in from all sides and maximizing the edge between our food forest and a wild system. For our forest garden, it took 15 youth-participants of the June 2011 Purple Thistle Institute 20 straight days to plant a massive sheet mulch and establish a winter cover crop of rye to build the soil and provide shelter and biomass for young trees. We've planted some hardhack (*Spiraea douglasii*), a native wetland plant and nitrogen fixer with medicinal leaves and beautiful pink flowers, and are thinking about establishing some cattail (an excellent year-round food source which houses remediative bacteria in its rhizomes) and some skunk cabbage (the root is an anti-spasmodic), as well as mulching heavily and inoculating with fungi. We've also planted some birch, hazelnut and valerian (all native plants that can build soils, tolerate harsh conditions and wet feet and produce food and medicine) as a windbreak and as the start of a transitional zone between the food forest and the wetland.

The results have been astounding — going in several months from zero-fertility to a soil that can hold water through more than ten days of withering urban heat, play host to over 20 kinds of fruiting fungi (so far), support a wide range of soil life and give life to a diversity of young perennials in the harshest of environmental conditions. Work is ongoing — learn more at radiclebeets.wordpress.com and come down to visit. We host regular work parties, potlucks and workshops in East Vancouver.

> *Adam Huggins is a collectivist, herbalist, musician, permacultural-ist and filmmaker living in East Vancouver, a sweater of the salt of the Great Ocean and a busybody garlic-peeling hand-processing dumpster-diving propagator of plants and emulsions. You can explore his work at sunfishmoonlight.wordpress.com.*

## Cottonwood Gardens: Creative Un-Planning as Community Repair

BY OLIVER KELHAMMER

I spent a few hours recently, trying to help a group of volunteers identify some bamboo I had planted at Cottonwood Community Gardens in Vancouver some 20 years ago. The bamboo it turned out was vexingly hard to sort out. The individual plants had spread far beyond where I had originally put them and were now melded into towering groves, the picturesque stems clattering together in the fresh spring breeze and the grassy leaves rustling pleasantly overhead. Though taxonomically, it had become a bit of a mess it sure was a beautiful place.

One of the benefits I've noticed about getting older is that I am able, in my mind's eye, to travel backwards through time, revisiting things I have worked on over the years, peeling away the layers of accumulated experience to reveal the stages of how they came to be. Whenever I visit Cottonwood, with its groves of exotic trees, native wildlife plantings and carefully tended vegetable plots, I so clearly remember the desolate, trash-strewn wasteland I faced back in 1991, when I put in the first garden plot there. Amid tire ruts, whirling dust devils and the punctured, empty Lysol® cans from which terminal-stage alcoholics had sucked the last dregs, I picked away at the hard-baked earth. There weren't even earthworms in the ground in those days, though I did dig up a rusty pistol and a lot of broken glass. Through discarded syringes and a dead cat in a garbage bag I persisted, though I did at times feel a little foolish.

I was after all, a squatter, indignant at the recreational apartheid being waged in Vancouver's East Side by the pro-development Non Partisan Alliance who ran City Hall at that time, who, despite their name, were anything but nonpartisan. Under their reign, million dollar beach front properties were being acquired for parks in more afflu-ent neighborhoods on the city's west side (the NPA's traditional base of support), while back in East Van, we had to fight hard to get even teeter-totters replaced in our overcrowded playgrounds.

I also had gotten word that city engineers were planning to annex the strip of land, which was on the southern flank of Strathcona Park, to push through a highway, expressly designed to facilitate more trucks getting in and out of the downtown core. This plan, had it succeeded, would have marooned the park, the area's largest, inside a forbidding cordon of high-speed roadways. Nearby, the newly formed Strathcona Community Gardens, which I also had been involved with, had just lost half its land to a recently built senior's development. This loss of space caused an unprecedented waiting list for its remaining garden plots. So I felt something just had to be done and squatting seemed the most productive alternative. I hoped that if I struck a claim to that sad little field of dust and weeds, others might eventually join in, and we would pry the property loose from the hands of the city's traffic engineers and politicians and turn it into a neighborhood commons that we would design and build together.

This, amazingly enough, was what eventually happened. As I spent more time in my little plot, mulching it with dead leaves, sowing a cover crop of buckwheat and planting potatoes, I was visited by var-ious passersby, many of whom were keen to start their own garden plots, which I was more than happy to let them do, despite having no vested authority to do anything of the kind. Soon a ragtag collec-tive formed, a band of green guerrillas who came from many walks of life. There was Byron, an unemployed lounge musician, who visited every day with his little dog Elvis. There was Mrs. Yee, always laughing as she the shared oranges she'd glean from the waste bins of the produce wholesaler across the street. Austin and Chiang-Mai Yu, senior citizens from rural Guangdong Province, set about transforming a large section of the wasteland into a productive mini-farm, bringing in shopping cart loads full of discarded tofu and digging it into the inhospitable soil. Lawrence, a soft-spoken former farmer originally from Saskatchewan, scoured the city's alleyways and brought us useful building materials and knickknacks that he'd proudly display on an impromptu free table. Juan, emphatic and stocky, labored non-stop at excavating a pond, which we soon christened Juan Lake, to capture rainwater. And there was Jack who looked as if he'd stepped right out of the Old Testament. He lived, parked next to the garden, in a white minivan, and if I happened to pass by at night, I'd see him hunched up under the dome light with his ancient Underwood on his lap, tapping out stream-of-consciousness manifestos he'd hand to the random train workers he'd come across on his self-appointed rounds of the nearby rail yard. There several others too who joined me; colorful, kind, somewhat crazy people, all of whom shared somehow my optimism that what we were doing, despite the daunting challenges, was eventually going to work out.

To make a long story short, after an explosive public meeting at City Hall, we were miraculously granted a lease. To our surprise, some of the foes we'd had on council relented after they realized that the amount of trash we were picking up was saving the city some serious money. The planned highway in the meantime had also been put on hold thanks to budgetary constraints, so there were even less reasons to oppose us. Having gained security of tenure, we were able to accelerate our development of Cottonwood and the rest, as they say, is history.

But 20 years later, what does it all mean? Is it significant that a small group of low-income people, living in Canada's poorest neighborhood, were able to turn a vacant lot into a horticultural paradise? Most emphatically — yes! Why? Because in doing so, we fundamentally questioned the power relationships extant within the modern city. We needed to ask ourselves: who gets to design a city and why? Does giving the design of our public spaces over to so-called professionals always necessarily provide the best outcomes for people and the environment? How can we make cities sustainable and responsive to the needs of their inhabitants: all of the inhabitants — human and non-human — and not just the privileged elites?

Without any special training, our ever-changing and diverse little group figured out how to cross cultural and language barriers, raise money, remediate toxic soils and cultivate a myriad plant varieties, many the first of their kind to be grown in Canada. We installed irrigation systems, built greenhouses and produced massive amounts of compost from the waste we gleaned from the nearby produce wholesalers (waste which otherwise would have found its way to a landfill). Not that there weren't difficulties and even occasionally vehement differences of opinion, but through it all, Cottonwood continued moving forward as a thriving, self-repairing paradigm of urban sustainability.

Even if they are well meaning, attempts to plan the urban landscape from the top down tend to create sterile, *theme-park*-like solutions that fail to address the complexity of people and nature. We've all experienced these places: exsanguinated, bourgeois-inspired zones that owe more to Ikea than Arcadia and leave us with the unsettled feeling that something essential, something authentic is missing. Cottonwood is the opposite of that. It exists in a constant, generative state of *un-planning*, where everything is provisional and adaptive to the ever-changing needs of those who choose to engage with it. All that is for sure is that Cottonwood will never be finished and it will never stay the same. Such open-endedness is essential for neighborhoods and the cities in which they find themselves to be able to grow into more inclusive, more interactive and more equitable places. Most communities, no matter how marginalized, have it within to repair themselves and their ecological underpinnings through preexisting social networks. People, ultimately, care about the places they live in and will try hard to make things better. To unleash this amazing power, bureaucrats and politicians should provide what resources they can, but mostly they need to learn how to give up a little control and just step out of the way!

One area where the authorities could be of use, however, would be to make free soil tests available to anyone who wants to grow food in a city. Urban soils are frequently contaminated by heavy metals and hydrocarbons that settle onto the ground from air pollution or seep in from rubbish. In the 1980s I attempted, as a land art project, to decontaminate the front yard of a Toronto art gallery located in a neighborhood where much of the soil had been contaminated with lead from a nearby battery reprocessing factory. I sowed successive crops of buckwheat, which is known to absorb heavy metals, and I distributed pamphlets that detailed the injustice of a situation in which local people, predominantly working-class Italians and Portuguese, were no longer able to safely grow produce in their yards because of the actions of an unregulated polluter.

Fig. 5.7: *Early days at the Cottonwood Community Gardens.*
CREDIT: OLIVER KELHAMMER

Fig. 5.8: *Today at Cottonwood Community Gardens.*
CREDIT: OLIVER KELHAMMER

Though my effort was largely symbolic, it gave me some pointers for doing longer-term remediation projects such as at Cottonwood Gardens, where our initial soil testing revealed pockets of contamination within some areas of the property. Because readings on much of the site were relatively normal, we reserved the few dubious areas for non-food crops such as black locust (*Robinia pseudoacacia L.*) and timber bamboo (*Phyllostachys bambusoides*), with the idea that they would slowly sequester the heavy metals within their fibers. We would then harvest the materials for long-lasting garden trellises and fencing, thereby keeping the toxins out of the food chain. Our strategy for removing organic pollutants such as hydrocarbons was to gradually let them break down via the mycorrhizal activity we were encouraging through our frequent applications of wood chip mulch. Though these remediation processes are slow, the beneficial side effect of having a crop of useful, locally sourced building material is more than worth the wait and adds yet another output to our thriving urban commons.

*Oliver Kelhammer is a Canadian land artist, permaculture teacher, activist and writer. His botanical interventions and public art projects demonstrate nature's surprising ability to recover from damage. His work facilitates the processes of environmental regeneration by engaging the botanical and socio-political underpinnings of the landscape, taking such forms as small-scale urban eco-forestry, inner-city community agriculture and the restoration of eroded railway ravines. His process is essentially anti-monumental — as his interventions integrate into the ecological and cultural communities that form around them, his role as artist becomes increasingly obscured. He describes what he does as a kind of catalytic model making, which lives on as a vehicle for community empowerment while demonstrating methods of positive engagement with the global environmental crisis.*

## Gas Stations and Sunflower Dreams

Old gas station sites are among the many toxic sites that litter our urban and rural landscapes. Their leaking underground storage tanks, also referred to with the amazingly ironic acronym of LUSTs, contain hazardous liquids, primarily petroleum products such as gasoline, diesel, kerosene or oil. Many of these old tanks are leaking and forming hydrocarbon plumes that contaminate groundwater, drinking water and the surrounding soil. As of March 2012, over 504,000 leaks had been recorded from federally-regulated LUSTs in the USA.[10] The most hazardous components of petroleum products are the BTEX compounds: benzene, toluene, ethylbenzene and xylenes. Benzene is the most hazardous of these compounds and is a known carcinogen. Another toxic compound in gasoline is methyl tertiary butyl ether (MTBE). Besides hazardous and toxic compounds (e.g. benzene, MTBE), there are many other pollutants caused by gas leaks, runoff and spills, from heavy metals to solvents. Even the concrete pad of a gas station leaves a toxic legacy in the soil: asphalt can contain PCBs as a flame retardant.

When I first learned that you could use sunflowers to clean up toxic landscapes, every fenced off and abandoned gas station lot I dreamed into sunflower fields. I would walk by them at night and try to figure out how to get over or through the fence with buckets of compost and seed the parched toxic soil so that the healing could begin. A few years later, I was having lunch with a friend of mine who works for an environmental consulting company and I was telling her about my sunflower infatuation, to which she delivered a strong reality check. She explained to me what sort of pollution was actually found at an abandoned gas station site, as well as how deep the contamination could reach.

Contamination found at these sites, like the hydrocarbon plumes that contaminate groundwater, and the soil, can run deep and spread over an area that extends well beyond the border of the site itself. This makes remediation challenging because you can't always get at the contamination, especially if you are in a densely built and populated area where that hydrocarbon plume may have migrated and now extends under the road, houses, and other businesses. In situations like these, some higher-tech solutions try to aerate and encourage bioremediation under these areas, or off-gas the contamination. Other conventional strategies include fencing and sealing up the site or excavating the contaminated soil, trucking it off to a landfill and filling the site with "clean soil." Due to their concern over liabilities, companies that own these sites may not allow you to engage in grassroots bioremediation work there, especially if it involves digging in the soil.

Gas station rehabilitation requires much more than seed bombing sites with sunflowers. When it comes to using plants to clean up toxic sites, plants can clean up the soil only as far as their root systems reach.

Using fast growing trees, such as poplars, willows or black locust, to access deeper contamination with their longer roots would be ideal. Following that with more surface soil remediators would also be great, though the main focus would be on working with the trees. This approach would take time, and success would depend on how deep the contamination or the hydrocarbon plume was and where it was in relation to the site.

Some projects working with old gas station sites lay down some sort of cap of clean soil, wood chips or other protective barrier, and then use raised beds on the site to avoid digging into the ground at all. This allows the site to be utilized in a rather safe way, but does not deal with the underlying material which may continue contaminating water sources. A mixed approach involving some trees to engage in remediating the deeper contamination with raised beds (which would not be affected by the closed tree canopy in the early years) could be a good place to start.

## Food Gardens and the Dreaded Lead

Lead contamination in soils is a big concern for urban food growers and gardeners; it has been caused by lead-based paint, lead-based gasoline, coal combustion and other products. In cities, older neighborhoods tend to have higher amounts of lead contamination. Soil nearest to structures (especially those older than 1978) and roads have the highest concentrations. Lead can also be found in vacant lots where a building once stood. Below is some helpful information and tips on dealing with lead in the soil that offer safer ways to proceed for all you green thumbs.

First of all, it is important to test for lead in your soil and to figure out how much lead is there. With most contaminants, the different environmental regulatory bodies have safe concentration levels and unsafe concentration levels, above which that level of contaminant is considered a dangerous health risk. It is important to understand that these levels are not as objective as they seem. They are subject to political pressure and industry lobbying, which means that they may be set higher than is truly safe for you. And in the cases of many contaminants, no amount is considered good for you. These levels also do not take into account long-term exposure or the combination of that contaminant with other contaminants you may be exposed to, and the resulting health impacts. Nonetheless, these regulated exposure limits are the guidelines we have. In the case of lead, Canada's acceptable residential and parkland limit of lead is no higher that 140 ppm. The agricultural lead limit is 70 ppm, which is what you should observe if you plan on growing food on the site. Rather disturbingly, the US EPA's threshold is at 400 ppm. European countries have safe lead levels below Canada's (Sweden's is 80 ppm and Italy's is 100 ppm).[11]

It is important to remember that there are several different forms of lead compounds in our soil, some naturally occurring, others human-made. Some are toxic, some are not. Most soil tests measure the amount of total lead in your sample, which means that they do not tell you how much of the lead present is of the toxic or non-toxic kind. They also do not tell you how much of that lead is bioavailable, meaning that it is actually mobile in the soil and able to be taken up by plants.

Overall, if you are dealing with lead (or any other heavy metal) in your soil, take the following damage control and safety precautions to limit lead exposure to yourself and your family:

- Do not wear your outside/garden shoes or clothing indoors; they can bring soil contamination indoors.

Fig. 5.9. Credit: Oliver Kelhammer

- When working, wear gloves and tight-fitting face mask or respirator.
- Wash your hands after you garden and before you eat.
- Locate your veggie garden at least ten feet from heavily traveled roads and old painted structures (e.g. houses, building).
- Add mulch (leaves or wood chips) and plant ground cover/cover crops to stabilize the soil and reduce bare soil so that it doesn't get blown around by wind, splashed up by rain or ingested by children and animals. Sheet mulch would be a great option here: laying down cardboard and then layering your compost layers on top of that.
- Grow mostly fruiting crops. "Most of the lead that enters a plant accumulates in the roots and secondarily in the leaves, though there are some exceptions. Fruits such as tomatoes, peppers, melons, okra, apples, and oranges and seeds such as corn, peas, and beans generally have the lowest lead concentrations and are the safest portions of the respective plants to eat if grown in lead-contaminated soils."[12] Safest does not mean free of lead. Please remember that lead accumulation and location differs according to each plant species. Definitely avoid eating roots, stems or leaves of plants if your soil has high lead levels. Do not plant greens (e.g. broccoli, kale, mustard greens, spinach and lettuce); common greens may concentrate lead in their leaves. Cabbage is the safest of leafy crops.
- Discard old, outer leaves of vegetables before eating; always peel root crops; wash all produce thoroughly with one tablespoon vinegar per 1½ quarts of water.
- Grow your produce in raised beds for the least amount of lead contamination (as long as the soil you add to the raised beds is truly clean soil and is not contaminated with lead).[13]

Aside from the above precautionary actions, there are three main accessible grassroots bioremediation options available.

### Option #1: Immobilize Lead in the Soil

### Bind Lead by Raising Soil pH

If you want to keep lead from moving into plants, increase the alkalinity of your soil. You can do this by adding organic matter, amendments or compost that is more alkaline.

## Bind Lead by Adding Phosphorous or Iron

Adding phosphorous also decreases availability of lead to plants, as it binds to the lead and forms *pyromorphite* crystals, a form of lead which is non-toxic and not bioavailable. Some good sources of phosphorous are fish bones, bone meal (calcium phosphate), bat guano and chicken manure. Rock phosphate has been used more conventionally, but some folks prefer to stay away from it as it is mined. You can also add some of these amendments to your compost to create a phosphate-rich compost to add to your site. By lowering the pH of the soil to make the lead more mobile and adding a soluble form of phosphorous, the two can combine into *pyromorphite.*

Make sure you allow some time for the phosphorous and the lead to react in the soil. Then test your soil to see if pyromorphite is indeed being formed or not. If not, then you may have to play with different sources of phosphorous and the pH until you find a good match. Following that, you then bind the lead further by bringing the pH back to neutral, build the soil up using compost, sheet mulching or raised bed techniques or establish a vegetative cover over the site.

In South Prescott, an Oakland, California, neighborhood where the US EPA West Oakland Lead Project is taking place, fish bones (which contain the phosphate mineral apatite that seems to combine especially well with lead to form pyromorphite) are tilled into the soil needing remediation. The soil is then covered with some sort of green cap (sod, clean soil with mulch, plants, raised beds or gravel).[14]

In some extreme cases, the combination of lowering the pH and adding increasing phosphorous concentrations can have adverse impacts on soil microorganisms, so you may want to reinoculate later with compost tea and compost.[15] Lead immobilization using phosphorous has been tested on old mine and industrial sites. However, garden soils can be richer in phosphorous so adding too much phosphorous amendment can sometimes damage plants by causing iron chlorosis.[16] Determining how much amendment to add and its application rate can be tricky. Proceed with caution, start small before applying it to a larger area and consider soliciting some professional guidance.

If you are adding phosphorous, make sure you are not also dealing with arsenic contamination, as phosphorous actually enhances plant uptake of arsenic. To immobilize/bind arsenic in the soil, add an amendment that has iron. Some experts favor using organic amendments such as a compost that is high in iron over phosphorous when it comes to binding both lead and arsenic. Iron compounds bind with lead through adsorption, creating iron-based compounds like *ferrihydrite*. Sources of iron to enrich your compost or soil can come from adding iron rich plants, like nettles or dock, either into your compost or making it them into plant extracts. Greensand is also high in iron, but a less sustainable source as it has to be mined. Biosolids are sometimes used in conventional bioremediation projects, but as mentioned earlier in the chapter, they can be potential sources of contamination. I would recommend you do more research and know the source and composition of the biosolids before you decide to go that route.

### Option #2: Extract Lead from the Soil

If you want to facilitate lead being drawn out of soil by plants, then make your soil more acidic. You can do this by adding more acidic compost or amendments (sulphur, coffee grounds, pine needles) to your soil. Do not add phosphorous-rich materials or compost as that can block extraction. Once you've done that, then plant multiple rotations of lead-accumulating plants (e.g. geraniums, sunflowers, brown mustard, alpine pennycress, vetiver), and be sure to take the proper precautions with their testing and disposal.

The more rotations of phytoextractors you use, the more lead you will withdraw, as long as you keep the soil at an optimal acidity for uptake. Some folks have found lead phytoextraction to be successful, while others have found it slow and challenging.

### Option #3: Combo Approach

Do a combination of a few rounds of phytoextraction with key plants, followed by trying to bind the remaining lead via organic matter, liming the soil or adding phosphorous. Then build the soil up using sheet mulch or raised beds so that plant roots are not drawing too much from the contaminated soil.

Finally, if you are dealing with dangerously high levels of lead contamination (the USA EPA says 2,000 ppm or more), consider forgoing phytoremediation and instead focus on really covering the exposed soil effectively. Start over with six inches of clean soil or compost, then cre-ate a barrier or find a way to stabilize the soil, the objective being that that there is no bare soil showing. For this, you can use thick mulch (wood chips, leaves) or landscaping fabric, crushed stone, patio or step-ping stones. From there, you build raised beds using clean soil or plant non-edible plants.

## Toxic Soil Busters Fight Back Against Lead

BY ASA NEEDLE AND MATT FEINSTEIN

The Toxic Soil Busters Co-op shows that improving a neighborhood's health doesn't have to be limited to experts and outsiders.

In 2001, a group of residents and students in Worcester, Massachusetts, USA came together to find a solution to a neighbor's high soil lead concentration that was contributing to their child's elevated blood lead levels. There were resources for removing lead inside the home, but what about the soil? At first glance, the solutions (namely excavation) seemed expensive and out of reach, but upon further research we were intrigued by lab studies that showed plants accumulating lead in soil.

We decided to educate ourselves. We started community workshops and then experimented in the field with the process called phytoreme-diation. Teams of neighbors were intrigued and energized by the tactic of using nature to heal itself. This was a creative, hands-on, visual and concrete approach. We wrote grants to cover testing costs to be able to offer free soil testing to residents. Many approached us with interest in gardening or making their yards safe for children. We then offered multiple options for getting soil lead levels under the EPA's safe limit of 400 parts per million. This included a combination of phytoreme-diation, building raised garden beds, bringing in clean soil and a variety of landscaping techniques.

We encourage organizations to experiment with phytoremediation in their own communities. You can plant scented geraniums or other hyperaccumulators like brown mustard, sunflowers or spinach. Plant as densely as possible in the affected area, and lead will accumulate in the roots, stems and leaves. After the growing season, safely dispose of the plants. If you put them into the municipal waste stream, ensure that your area has good lead protection on incinerators or landfills. Note that this technique is very slow, and depending on the lead concentrations and soil conditions, remediation can take several growing seasons.

In most cases, Toxic Soil Busters combines phytoextraction techniques with other lead-safe landscaping techniques. Stabilize the soil by planting plants that grow soil-retaining roots systems such as shrubs or ground covers to reduce foot-traffic access and dust creation. This process is called *biostabilization.*

When we formalized the organization and made a strategic vision for this project, our first task was to define *environmental justice.* We learned quickly that it is not static word, but rather a movement that is ever changing. For us, it is beyond land conservation and more about the spaces where people live, work, play and grow. It is about how toxic our backyards are, how little green space our neighborhood has, how unhealthy our homes are and how poorly the public transportation system works for people on our street. The justice aspect lies in how these health and safety deficits disproportionately affect low-income folks and people of color. Maps showing where lead-poisoned children live revealed a glaring concentration in our inner-city neighborhoods. We had found our issue!

Toxic Soil Busters Co-op has become part of a movement to create a new, green economy based on sustainability and solidarity. The co-op equips youth with skills that are in demand, and yet allows them to focus on ethical livelihoods. Implementing a model of deliberative democracy, we use a horizontal decision-making process to arrive at consensus. We are dedicated to transparency, equality and good business practices, providing jobs and training to at-risk youth. The wisdom of the group is a powerful resource, and the business profits from this open trafficking of ideas. Youth are more motivated to work hard and come up with creative solutions when they have a stake in the future of the collective, when they feel they own a piece of it. We exist as evidence that youth can run democratic worker co-operatives. Youth are not merely victims in need of assistance, but are powerful forces for positive transformation. We believe that only youth can know how they can best reach out to their peers. They can be community leaders,

movers, shakers and educators. Youth are not the leaders of some uto-
pian future. They are the change agents of today.

Fig. 5.10: *The Toxic Soil Busters.* CREDIT: ASA NEEDLE

*Asa Needle is Outreach and Education Coordinator of the Toxic
Soil Busters Co-op. Matt Feinstein is the Co-Director, Media and
Organizing Coordinator of Worcester Roots Project's staff collective.*

For more information about the work of the Toxic Soil Busters
Co-op, check out worcesterroots.org/projects-and-programs/
toxic-soil-busters-co-op/

To download their *Lead Safe Yard Manual: A Do-It-Your-Self
Guide to Low-Cost Soil Remediation and Safe Gardening* go to
worcesterroots.org/2011/08/15/lead-safe-yard-manual/.

## GTECH Sunflower Gardens —
## Growth Through Energy and Community Health

Located in Pittsburg, Pennsylvania, USA, Growth Through Energy and Community Health (GTECH) Strategies is a nonprofit social enterprise which invests in community revitalization through green economic development initiatives. They partner with community organizations, service providers and public agencies to develop and implement community-based projects that include vacant land transformation, green infrastructure, weatherization and retrofitting, deconstruction, residual waste reuse and biofuel production. Vacant land is plentiful in the city of Pittsburgh due to steady population decline and economic downturn since the 1950s. Most recent calculations indicate over 25,000 vacant parcels of land, and the problem continues to grow. Many of these vacant lots are subject to littering and dumping garbage, leading to a whole spectrum of soil contamination. GTECH's Urban Revitalization Program works with community groups to assess, plan and implement transitional strategies, like sunflower gardens, while working toward a sustainable, healthy and long-term productive reuse for each site.

GTECH's sunflower garden projects are a mix of phytoremediation, biofuel production and community revitalization. In an interview, Andrew Butcher of GTECH explained to me how he had gotten really excited about pairing up alternative energy and community development by planting biodiesel crops on vacant land, to improve soil quality, create biofuels and green jobs. According to Butcher, the sunflower gardens are a high-impact, low-cost transition strategy to invigorate communities while helping improve the state of the vacant sites. "It is so visual and compelling to see a blighted lot transformed first into a place bustling with sunflowers and then into a place with green infrastructure, like a community garden, park or food forest. Sunflowers and their phytoremediation capabilities are an added benefit to the powerful visual they create and their seeds are an added benefit for the biofuel process. It's a catalyst for greater transformation," said Butcher.

In its short five-year life span, GTECH has worked with over 40 community partners and with hundreds of community sites. In that time, about one third of their sites have transitioned to become parks, urban farms and community gardens, playgrounds and food forests. They have worked with community groups to find funding to employ local residents in the maintenance of the sites, as well as partner up with other organizations to provide training programs.

On each site, GTECH conducts extensive soil testing to determine what levels of contamination are present. They also test the tissues of the sunflowers at the end of the season to determine contaminant uptake so that they can decide whether they should take the harvested biomass back to a composting facility they have partnered with so that it can be mulched and composted (if no toxins are present) or if it has to trucked away as toxic waste.

Like many phytoremediation projects, the promise and potential are inspiring, but the on-the-ground reality can be more challenging. Initially the whole idea of using the sunflowers to yield biofuel was a key driver in the strategy, even though GTECH knew that oil yield would be minimal because they were growing crops in poor soils. Not all sunflowers yield large amounts of oil from their seeds. There are specific oil seed sunflowers, which are lower to the ground and smaller. GTECH found that using these sunflowers radically reduced the visual impact of the project and did not block enough sunlight to understory weeds. Also, harvesting the seeds, crushing them and extracting the oil involves some pretty expensive equipment. In the end, communities preferred larger plants and fewer weeds.

So GTECH's efforts shifted to using these gardens as demonstration and education sites and reactivating these spaces. Instead of making biofuel from the sites, they save many seeds for replanting and for sale. Other sources of excess, waste cooking oil in Pittsburgh's urban core can be more easily used to make biodiesel. According to Butcher, community biofuel production, phytoremediation and green jobs are all great ideas to incorporate into a larger good for the planet good for people strategy, but each has its own challenges. Therefore, there is need to identify creative solutions and partnerships to figure out how to push all the edges from the realm of the impossible into the possible.

Fig. 5.11: *GTECH project site.* CREDIT: GTECH

Most GTECH projects revolve around small-scale community sites, not large-scale brownfields. As a result, getting access to the sites can be much easier and often quicker than working with companies who own the larger contaminated sites. GTECH also believes that not every vacant lot will become a green space in the long term. In some cases it makes more sense for a site to transition into affordable housing or some sort of local economic development projects instead. "Green groups focus on green outcomes, community groups focus on community outcomes. We focus on the intersection between the two," added Butcher.

Each site is different when it comes to phytoremediation, and GTECH often deals with a whole spectrum of contaminants. "GTECH hasn't found the key to unlocking phytoremediation. We contribute to improving the remediation process, but phytoremediation is a long process and there is no one size fits all, so we just need to be realistic," said Butcher. Phytoremediation can be hard to get started effectively with minimal resources on degraded sites, but half the battle is building soil quality. To address soil quality issues, GTECH has been adding large volumes of compost, and they are constantly looking at other ways to build value into that soil. GTECH also has a partnership with the University of Pittsburgh to research the practicality of their remediation strategies. However, they are finding that even with a university partnership, a lot of money is required to procure the resources needed to adequately test their phytoremediation efforts.

Challenges aside, the projects are still a lovely and inspiring success. "Soil quality on our sites is better after our projects than before. With more organic matter on-site, there is the ability to lock the contaminants in the soil or dilute them, and this is where compost really comes in ... copious amounts of compost. Sunflowers do pull out some contamination and compost really builds the soil and its ability to deal with the toxins. Every site we are on has an exponentially better chance to be transitioned into agriculture or green space. Action really begets capacity, and capacity and knowledge really begets long-term impacts and changes places," Butcher stated.[17]

*For more information about* **GTECH** *and the work they do, please check out their website gtechstrategies.org.*

## It's Time to Phytoremediate!

Hopefully after reading all this, you feel more empowered to use phytoremediation as one of your earth repair tools. A foray into phytoremediation will involve hard work and time. It will be relatively cheap and likely yield lush and dynamic healing for the Earth and community you are working with. You may be starting out with more questions than answers, so proceed with curiosity, creativity and determination. Soil and plant testing are key to assessing the success of your efforts. Combine phytoremediation with microbial and mycoremediation for maximum effectiveness. Share what you learn with others — this is an area of healing work where we need more local information and experimentation! It all begins with you taking the first step, so dig in.

CHAPTER 6

# Mycoremediation

THIS CHAPTER WAS COMPILED from the contributions of several different grassroots mycoremediators, as well as interviews with mycorestoration expert Paul Stamets, Mia Rose Maltz from the Amazon Mycorenewal Project and Scott Koch of the Telluride Mushroom Festival.

## Fungal Remediation and Radical Mycology

BY PETER McCOY

### Fungal Remediation

The more one studies the fungal kingdom, the more it becomes clear how little is actually known about it. Indeed, we are only beginning to understand the discoveries that have been made in recent years regarding the abilities of certain fungi to reduce the impacts of pollution and industrial disasters. This research in the field of fungal remediation has developed a wide variety of applications using certain fungi for environmental cleanup, from water filtration and soil cleanup to erosion control and heavy metal sequestration. With roots found in several decades of research by countless scientists around the globe, these amazing advances have in recent years finally found their way into the public sphere. These techniques are now frequently referred to by the popular terms *mycoremediation* and *mycorestoration*, which encapsulate the various restorative roles that fungi serve. Increasing public awareness of the remediative fungi has even begun to expand into a veritable mycological movement that will only continue to grow as people around the globe expand upon and refine these practices.

Fungi are found in every corner of the planet, from the permanently frozen tundras of the north to the heart of Antarctica. This branch on the tree of life includes species that form a variety of symbioses with plants, others that facilitate succession in the forest canopy by killing off weakened trees and still others that recycle the organic matter of the world. Found in our stomachs and on our skin, fungi are integral to essentially all natural life cycles, and their incredible diversity in the world deserves all the respect they are given.

However, it is to the saprophytic fungi that we turn in our search for remediation potential. *Saprophytic* is a general term used to describe any fungus that plays a role in the decomposition process of matter; thus they are also known as the *decomposing fungi*. These fungi and the powerful digestive enzymes they produce have been studied in recent decades for their respective abilities to break down or reduce a wide variety of chemical and biological contaminants. The saprophytes include many well-known and commercially cultivated edible and medicinal species (e.g. shiitake, oyster, portobello and reishi mushrooms). Before we jump right in to working with these fungi in remediation, however, I will first go over some basic background information so you best understand how these amazing beings do what they do so well.

## The Basidiomycete Life Cycle

The fungal kingdom is divided into several subgroups, or *phyla*, each with its own unique life cycle. One of these groups, known as the *basidiomycetes*, includes all of the species you are likely to try cultivating for remediative purposes (e.g. oysters). I recommend that you come to understand the saprophytic basidiomycete life cycle before you begin cultivating fungi for remediation. Understanding the fungi's life cycles will certainly help you to not only understand what aspects of nature you are trying to mimic throughout your cultivation trials but will also (hopefully) result in greater successes and fewer contaminations from competitor molds and bacteria that are trying to eat the food you are providing for the mushrooms.

We begin with the spore. *Spores* prolifically develop on a microscopic layer of fertile (spore-producing) tissue known as the *hymenium*. This tissue develops in mature mushrooms on the surface of structures called gills, teeth or pores, which themselves are often found underneath the cap of a mushroom. A given mature mushroom can produces millions, or even billions, of spores in a single day, all of which the mushrooom ejects at an incredibly high force to enter their surrounding environment. When a given spore lands in a suitable habitat, it quickly germinates, producing a single-cell filament, or *hypha* (plural *hyphae*), which begins to grow through its *substrate* (food source) in search of a genetic mate. Like the sperms and eggs of animals, spores contain only half the genetic information of their parent and thus need to join with the hypha of another spore in order to be genetically whole.

Once the spore does encounter a mate, the two hyphae fuse into a joined network, which is then referred to as *mycelium*. This mycelium now has all the genetic information it needs to successfully grow through its environment and ultimately produce mushrooms. As the mycelium grows through its substrate, this thread-like structure continuously branches in all directions, forming an incredibly dense network (imagine a web with clearances smaller than any woven structure humans can produce) in the search for water and food. In the case of the saprophytes, as the mycelial tips encounter organic matter a mixture of complex enzymes is exuded which converts this material into forms the fungus can use as food. The main energy source for these fungi is the long chain-like molecule of *cellulose* (the fibrous stuff that makes up the walls of plant cells). Saprophytes have developed an array of enzymes that can readily snip this long chain in to simpler, shorter carbohydrates that the fungus can then absorb and metabolize. Some saprophytes have even adapted to break down *lignin*, the highly complex compound that makes wood hard and rigid; this is something few things on Earth are able to accomplish. As the fungus is producing these degrading enzymes it is also releasing various metabolites to protect itself from competitors in the surrounding environment. Being only one cell thick, the mycelium has no outer barrier to infection and thus has evolved to defend itself from harmful bacteria and fungi in its substrate through the use of its own antibiotics and antifungals.

If the fungus runs out of resources or a change in environmental conditions arise (e.g. a temperature drop and increase in humidity), the mycelium will be triggered to produce a mushroom (that is, to *fruit*); it will start to accumulate into numerous tiny pinheads, or *primordia*. After a few days, these primordia will soon develop into a mature fruiting bodies (what we commonly refer to as mushrooms), at which point they will begin to drop millions of spores and continue the life cycle anew.

As noted above, one can think of spores as analogous to the sperm and egg in mammals in that each contains only half of the genetic information of its parent. However, unlike sperm and eggs, spores are not limited to simply seeking a single "opposite sex." Instead, some species have tens of thousands of combination possibilities due to the way the genetic information is shuffled when spores are formed on the hymenium. This means that, due to such a large number of possible genetic expressions, these fungi have an exceptional ability to adapt to a wide variety of environments and food sources. Thus a remediator can potentially "train" a given species to become accustomed to digesting a given pollutant. This training process will be explained in greater depth later in this article.

## Fungi in Remediation

The use of saprophytic fungi to heal the land can take many forms. Three of these abilities are of particular interest to those seeking to deal with large disaster scenarios: their functions as chemical destroyers, heavy metal chelators and ultra-fine water filters. We can use other remediative properties of fungi — such as pH correction, soil building and the intentional use of certain species to improve plant health — as techniques to help the planet. However, as *Earth Repair* is a book focused on disaster response, I will refer you later to the recommended resources for further study of these techniques.

As noted above, the saprophytes are incredibly powerful decomposers. They are responsible for around 90 percent of all decomposition in the world.[1] Without their efforts the world would be literally piled high in trees because saprophytic fungi are almost solely responsible for the breakdown of all the woody material in the world. As it turns out, the same enzymes that fungi have developed over millions of years to break the chemical bonds in cellulose and lignin have, in recent decades, been shown to also degrade many toxic and highly persistent chemicals. This is possible because the nature of the chemical bonds in cellulose and, for example, a pesticide molecule are relatively similar. When the fungus applies its digestive enzymes to the pesticide, the bonds in the pesticide are broken in the same way that they are in cellulose; both materials yield similar sugars that the fungus can readily metabolize. The mushroom thus can turn a large, complex chemical into many smaller, safer ones, theoretically ridding (or at least reducing) the surrounding environment of a pesticide's negative effects. This ability has been demonstrated with numerous fungi on a range of chemical pollutants. Notable examples include DDT, TNT, many petroleum products (including diesel, tar, motor oil and herbicides), chemical dyes, dioxins, PAHs and PCBs. In theory, the mushroom would still be safe to eat after this process. However, many mushrooms are known to hyperaccumulate heavy metals into their fruiting bodies. These metals, which are unsafe to eat, are found in many petroleum and industrial products. Thus, fruiting bodies formed on contaminated sites (or even on roadsides) should be disposed of properly and not consumed.

This ability to *sequester* (concentrate) heavy metals is another powerful role of the fungi that the DIY remediator may be able to utilize. Currently around two dozen species of fungus have been shown to aggregate either cadmium, radioactive cesium, arsenic, mercury, copper, lead or a combination of these metals. For reasons unknown, the mycelium works over time to channel these metals from its surrounding substrate into its fruiting bodies, which can then be removed from the environment for safe disposal. This amazing ability is still being studied, and in some situations seems to work better with *mycorrhizal fungi*. This type of fungus forms a symbiosis with plants and other microorganisms at the root level by extending the plants root system in exchange for photosynthesized sugars. Unfortunately, these species are not as easy to cultivate for even professional mycologists due the complex nature of this multifaceted interaction.

The third notable role fungi play in remediation is as an extremely dense water filter. As stated, the mycelium produced by fungi forms an extremely fine network on a microscopic scale. If a straw bale or burlap sack that has been fully colonized by mycelium is submerged in a contaminated waterway, the mycelium can work as a sieve to trap some of the passing pollution or biological hazards and thereafter degrade them as it does on land. This technique of fungal filtration is one of the most accessible to the DIY remediator as it is one of the easier techniques to implement.

## The Fungal Movement. Its Past, Future and Hurdles

From the tribes of northern Siberia to the philosophers of ancient Greece, fungi have played an important role in the development of many societies around the world. In Japan, mushrooms like shiitake have been cultivated on outdoor logs since at least the second century AD. And since the common button mushroom (*Agaricus bisporus*) was first cultivated in French limestone caves in the 1700s, the practice of eating mushrooms has been slowly creeping westward. Despite this long history with fungal consumption and identification, the contemporary science of advanced mycology really only goes back to around the 1920s when sterile laboratory techniques were developed that enabled fungi to be studied in a consistent and controlled manner. Since that time, some professional work with fungi has focused on the improvement of efficient (and profitable) cultivation techniques of edible and medicinal species. It has really only been in just the last few decades that interest has been paid to the abilities of fungi to help remediate polluted environments. And things haven't been the same since.

While the books like *Fungi in Bioremediation* by G.M. Gadd[2] and *Mycoremediation* by Harbhajan Singh[3] both provide excellent summaries of the scientific literature on fungal remediation, the more accessible book *Mycelium Running* by Paul Stamets[4] has drawn much attention to the subject in popular culture. With the help of *Mycelium Running* and online communities of cultivators, homesteaders and farmers have come to ally with the fungi to improve their gardens and land and realize the power of fungi to enact positive ecological change. At the same time other groups have taken it upon themselves to use these remediative techniques in their communities and other parts of the globe in search of healing and justice.

One of the most notable projects of this kind comes from a group based out of the Ecuadorian Amazon Basin. The Amazon Mycorenewal Project (AMP) is a volunteer-based organization working with fungi (mostly oyster mushrooms, as well as local Ecuadorian species) and bacteria to begin remediation on the 1,000+ toxic pools of crude oil scattered throughout the Ecuadorian jungle.[5] These pools, formed over decades of terrible oil extraction practices by Texaco and Chevron, have sat untreated for over 30 years, essentially ignored by the responsible oil giants and local governments. Through their groundbreaking work, AMP stands as one of the only current examples of independent, on-site fungal remediation work to be found outside of the professional world. Despite the repeated efforts by government to destroy their field tests, the project has not only kept strong for the last six years but has provided various outreach aspects to their work including classes for local communities on the use of fungi for remediation. AMP plans to be active into the foreseeable future and is always looking for interns and volunteers for their frequent work sessions.

While the work of AMP is truly inspiring, the sad fact remains that such remediative work with fungi is largely outside the sphere of public discussion and thus only slowly becoming recognized. If a person does become interested in learning more, however, the second barrier to advancing with this work is lack of accessible information and resources needed to study fungal remediation. Those in the USA seeking more advanced information on the topic are faced with few upper level courses in mycology and literature that is intimidating in its technical and, at times, vague nature. The problem compounds when you take in to account that private firms, large bioremediation corporations, government departments, contractors and universities have patented many currently known fungal remediation techniques. These patented methods are not available to the public and are not described in any readily available books. And yet, if a person were to apply these techniques to clean up their own land, they might unknowingly be infringing copyright through "theft of intellectual property." Those found guilty may be required to pay to have the site surveyed, permitted and remediated by a "professional," to the tune of thousands, if not tens of thousands, of dollars.

# The Radical Mycology Convergence (RMC)

At the time of writing, the second RMC has just wrapped up. Like the first in 2011, the 2012 RMC was unlike any other fungally focused gathering the world has seen. As a full weekend of free, low-tech fungal cultivation and remediation workshops, the RMC was an empowering (and fun) event for all who attended. Individuals came from various countries to impart their knowledge and meet like-minded people with whom they could share their passion for fungi and Earth stewardship. Through the RMC, organizers sought to develop a community of supportive and encouraging peers who could work with each other in a collective search for a better understanding of our fungal allies. In a world where finding accessible information on mycology is at times a test of endurance, many people left saying they had been waiting their whole lives for something like it.

A big take-away message from the RMCs is that there is so much yet to be discovered about mycology — and so few people doing it — that it will take the work and insights of amateurs to increase understanding. At the second RMC the scope of workshops was expanded to include microbiology and soil science as well as discussions on the cultural and philosophical role of fungi in our lives. At both RMCs, remediation experiments were installed to both reduce the impact of the event on the host location and to mitigate future human contamination of the land. Sample projects from both years include erosion control beds and greywater mycofiltration installations.

The plan for the coming years of the RM movement includes more international convergences as well regional gatherings. Future RMCs will be sure to include a solid foundational course in cultivation and remediation techniques, as well as related topics such as mushroom medicinals and foraging skills. The website[6] plans to develop into an online database of free resources on remediation information as well as a networking hub for the evolution of this network of radical mycologists. Many in this movement feel that it will only be through such grassroots efforts and community supported mycological work that the science of fungal remediation will move forward and become truly accessible. We are open to all who appreciate the power of the fungi and wish to learn more. Join us!

Clearly, proprietary information is not the way toward a freer, healthier planet. The answer is clear: we need more community-scale mycological remediation work and experimentation and, most importantly, transparent communication and collaboration between experimenters. Thankfully, a current movement is building around the world known simply as Radical Mycology (RM). A challenge to the closed-door world of professional remediation, the RM movement seeks to put the power of fungi in to the hands of the people, offering transparent access to the in-depth skills and techniques of fungal remediation. The RM movement seeks to destigmatize the fungal kingdom and show others how humanity can benefit from its many applications and, most importantly, how we can do it all ourselves. Thus, in the fall of 2011 the first ever Radical Mycology Convergence was held in northern Washington State.

## How to Cultivate

Using fungi for remediation purposes requires a large amount of mycelium. And, really, the more the better, as mycelium is what does the bulk of the important clean-up work. Acquiring this mycelium is the greatest challenge to the DIY remediator as an accessible abundance is not readily found in nature. Thus, the choice becomes either to buy commercially cultivated mycelium (a rather expensive route to take) or to learn how to grow your own. I hope that the following section will set you on the path toward the latter option. As daunting as cultivation may or may not sound, the good thing to know is that growing just the mycelium is the easier part of any mushroom-growing business. It's harder to provide the precise conditions needed to induce fruiting in the mushroom. Luckily for remediation purposes, we are not concerned with getting edible mushrooms but with applying large amounts of mycelium to our damaged environment. In the following sections, I have chosen to highlight some cheaper and simpler techniques to give you a taste of the numerous options available to the home cultivator. The skills below should be adequate to get you on your way toward success as you gain experience and confidence in your work with fungi. But, like most things, there is definitely much more to learn should you wish to deepen your cultivation skill set. For more options and ideas, be sure to check the recommended resources at the end of this article.

# A Note on Sterility

The perceived need for impeccable *sterility* and thus an impossibly clean and expensive workplace is one of the most common mental obstacles people encounter when thinking about fungal cultivation. While historically such strict practices have been the case — and still remain the standard for professional work — recent discoveries and boundary-pushing tests by amateurs and professionals alike have resulted in numerous low-tech and lower-risk processes for the home cultivator.

Sterility is of concern to fungal cultivators because without it, your cultures and experiments have a higher likelihood of becoming contaminated by competitor fungi and bacteria that are present in the air and on the surface of all objects. The competitors' ensuing feast may force you to throw out an entire project batch in order to prevent contamination spreading to the rest of your workplace. The techniques below deal with this legitimate concern by focusing on the use of "stronger" fungi that are able to defend against competitors while simultaneously creating environments that do not allow competitors to easily thrive. Also, these techniques below are considered *semi-sterile* and still deal with the threat of contamination to a fair degree while not requiring the expensive equipment of a professional lab (such as a $500–2000 air filtering flow hood) or precision in technique.

While the techniques outlined here have been successful in the outdoors with minimal precautions taken, people that find themselves prone to contamination may choose to follow these instructions in order to keep the workplace and cultivator (that means you) as clean as possible:

- Take a shower and put on clean clothes and a hairnet prior to working.
- Wipe all surfaces down with 80% isopropyl alcohol.
- Spray the air with Lysol® or 10% bleach then let the mist drop out of the air (theoretically carrying spores) several minutes before working.
- Use clean tools that have been dipped in isopropyl alcohol and then passed through a flame to achieve sterilization.
- Do not talk or open your mouth while working.
- Wear latex or nitrile gloves and frequently rub your hands with isopropyl alcohol to keep them clean.

The basic flow of growing fungi for the DIY remediator follows.

## Step 1: Acquire Source Spawn

*Spawn* refers to a bulk quantity of mycelium grown on any carrier substrate. *Source spawn* simply refers to your initial source of mushroom mycelium; it could come from any of the sources listed below. This source spawn is what will lay the foundation for your cultivation, as it will be used to grow your future mycelial stock.

To ensure the mycelium takes off fast and stays healthy, keep a few considerations in mind. While there is some flexibility allowed depending on your situation, do not ignore the following aspects if you are seeking the best results possible.

### TEMPERATURE

All fungi have a range of temperatures within which they can survive, with an even narrower range being ideal for fast colonization. These ideal temperatures vary between species though in general room temperatures should be adequate to get fairly good colonization rates. Too much heat can kill the fungus as can freezing conditions.

### HUMIDITY

Most cultivated mushrooms need some amount of ambient humidity to grow. If the food you provide is adequately saturated and the mushroom is not put in direct sunlight to dry out then it should be OK. However, ideal conditions would maintain heightened levels of humidity through misting/watering the mycelium as it dries out, keeping it in a terrarium or under a loosely draped plastic bag and providing shading to reduce the drying effects of sunlight.

### AIR

Like animals, all fungi breathe in $O_2$ and expel denser $CO_2$. Without a fresh supply of air the mushroom may not grow well. Further, stagnant air is an ideal condition for certain contaminants to flourish. For these reasons, always consider the availability of air to the fungus in all stages of its development.

## STRAIN

Many are not aware that, when attempting fungal remediation, not only may one species of fungus be best suited to your project but also there may be better-suited strains available to you. In the same way that the plant known as "tomato" has many cultivars that, while considered the same plant, grow and perform differently, so too are there numerous types (*strains*) of each fungal species. This variation originates in the thousands of potential genetic combination options available through the basidiomycete life cycle. Acquiring a given strain of a particular fungal species will determine several factors in your remediation work and is worth considering. For example, certain strains may grow faster and remediate more effectively on a given pollutant than others. Some strains may never form fruiting bodies (a potentially desirable effect). And native strains may be better adapted to the local environmental conditions and suffer less from competition. Finding the best strain for your project is a matter of luck in most cases, as only through trial and error will you be able to discover what works well for your situation. However, there is the possibility of developing a strain to work for your project, which I explore later in this article.

These considerations are not meant to discourage your efforts but rather to show that difficulties or unfavorable results may have multiple causes. That said, the following are good resources for source spawn.

## COMMERCIAL SPAWN

The simplest solution for sourcing spawn would be to purchase some amount of mycelium from a professional lab or local mushroom farm. The price tag attached to this option can quickly add up depending on the scope of your project and budget, but requesting donations or sponsorships might help reduce such costs. A potentially better option would be acquiring so-called spent kits from growers for free. *Spent kits* are blocks or bags of mycelium that have already produced a flush of mushrooms, which the grower has since harvested and sold. While these kits still contain very active mycelium (and may very well produce more mushrooms in time) as well as an abundance of digestive enzymes, these kits are not considered economically important enough to keep taking up space up in the fruiting room and are thus discarded to the compost pile. I imagine relationships could easily be forged between cultivators and remediators to get this valuable spawn to important projects if agreements between parties were made to the effect that the users would not profit from them. Spent *pleurotus* kits (aka oyster species, highly useful to the remediator) are often abundantly available from growers.

If you aren't able to strike up such a deal, the next option would be to buy your source mycelium. You can order kits online from multiple vendors if none are in your area. Even just buying one kit (roughly US$20–40) will get you well on your way to quickly bulking up your own mycelium supply. The benefit of buying from a grower is the likelihood that the strain they use is a strong and rapid grower. However, commercial purchase does not ensure that the strain you get will be as quick to colonize your polluted and less-inviting substrate. The big drawback here (apart from cost) is that the mycelium will in some sense be weakened after having gone through its life in a highly sterile environment. This means that the mycelium you buy may or may not equal a wild strain's ability to fight off bacteria and other fungi; it may be more susceptible to attack from ambient competitors on-site and/ or be less suited to the local climate. This is less of a concern with the *Pleurotus* (oyster mushroom) and other aggressive species, but it is something to consider in all purchasing decisions.

## MUSHROOM PLUGS

Plugs are roughly one-inch pieces of birch dowel that have been inoculated with mycelium. These are special dowels in that they have been cut with a spiral groove to give the mycelium a place to nest. You can buy inoculated plugs in bulk or in small packages from growers or online. Many species are grown on plugs. Depending on the vendor, you might be able to find various oyster species, turkey tails, elm oysters and shaggy manes. While buying plugs is a cheaper option than buying kits, you get a significantly smaller amount of spawn with this option. Thus, the savings made in money will be lost to time spent waiting to expand this smaller amount of source spawn. The same general benefits and drawbacks of commercial spawn strains apply to plugs as well.

## A Note of Caution

*Earth Repair* is not meant to give advice on the harvesting or identification of wild mushrooms. Be sure to learn these skills prior to foraging, and be sure to cross-reference your identifications with multiple guidebooks. Better yet, go with someone who has experience and use multiple guidebooks together.

### WILD / STORE HARVESTING

This approach is a great option for the total novice in that investments of time and money are next to zero, while the potential outcomes could set you up for the rest of your time as a mycologist. Picking wild mushrooms or selecting a few fresh and healthy looking specimens from a store is an easy way to gather a variety of strains cheaply.

If you want to find wild specimens, you will be glad to know that many of the species listed at the end of this article grow in much of North America and are relatively easy to identify and find. Even if you don't live next to a pristine old growth forest, you'll be surprised by how abundant mushrooms are at your local park at the right time of year (typically the fall for most species). If picking wild mushrooms, be sure to harvest the bottom of the mushroom body, where the stem interfaces with the ground and its mycelial network. This is where the mycelium is most active and viable for low-tech cultivating.

At the store your options are more limited. Here the most interesting candidate for remediation will likely be commercial oyster species. Whether acquired from the store or the woods, you will want to transfer the mushroom home in a moist paper (not plastic) bag and follow one of the bulking techniques below. Getting your mycelium sourced this way is easy and simple and can thus liberate the experimenter from any potential disappointment should things not work out. Ideally the mushroom you source would be young (and thus still actively growing) and as clean as possible.

### SPORE PRINTS

Spore prints and spore-filled syringes are available for sale online (US$5–15) but are generally intended more for sterile work which is beyond the scope of this book. A better option for the beginner cultivator would be to try making a *spore slurry* from mature, wild harvested mushrooms. This cheap and simple mycelium rich liquid, once produced, can be poured over your intended substrate in a survival-of-the-fittest approach towards developing a successful strain.

To try this technique you will first need to acquire several mature mushrooms of your desired species and then place them into a five-gallon bucket filled with non-chlorinated water, one to two grams (½ to 1 teaspoon) of table salt and 50 mL (¼ cup) of molasses. Tap water can be left out in the open for 24 hours to allow the chlorine to evaporate. Let the mushrooms sit in this broth for four to eight hours, during which time thousands of spores will be released into the favorable liquid. Remove the mushrooms and then allow this mixture to sit undisturbed for 24–48 hours at 50–80°F. The sugars in the liquid will soon stimulate the spores to germinate, sending out their microscopic hyphae. After a day or two, you may be able to see a thin layer of mycelium growing in the broth. At this point you can pour this spore slurry over your desired substrate, crossing your fingers that it will take hold.

From the thousands of genetically different hyphae that will grow, only a few will prove capable of finding a mate and forming mycelium that will eventually colonize the material. The strain(s) that has survived will thus be best suited for growing on your substrate and will continue on to colonize your material, providing you with a large amount of mycelium in time.

Taking this technique a step further, if this mycelium results in fruiting bodies, the slurry creation process can be repeated with these second generation fruits to create a new batch of spore slurry. By

You can try starting with 100 percent contaminated substrates for this method, but this likely will not work. The mycelium often seems to need more easily digested food sources to consume as it is remediating. Thus, mixing in some pasteurized straw or other organic food for the fungus is preferable. Also, by starting off with a lower concentration of contaminated substrate to fresh substrate in the first spore slurry generation and thereafter increasing the ratio of pollutant to fresh material with each generation, your success with "training" will likely improve.

applying this new spore slurry to your substrate you will be producing a new, and perhaps even better adapted, strain. In essence, you are "training" the fungus to cope with a given environment and substrate. If repeated further times, every successive generation will (in theory) be better adapted than the last at dealing with the contaminants you are working to remove.

### Step 2: Bulk Your Spawn

Now that you have some amount of mycelium, you will need to expand it in to *bulk spawn*. The techniques below are good starting points for beginning cultivators to bulk up, or expand, their spawn and should result in a fair degree of success if followed closely. Be aware that this is a stage where hopes for quick successes can be easily crushed if you set expectations too high and develop sloppy habits. To avoid these pitfalls be sure to not get too far ahead of yourself. Follow the techniques listed here before moving on to more advanced work. This will help you to build an important foundation of experience as well as confidence that can lead to more technical work in the future. You should realize up front that many people have problems with contamination or slow/no mycelial growth their first few times cultivating fungi. But if you go into the process knowing this fact and are ready to face some failure, you will be that much more resilient when the cold green face of *Trichoderma* mold stares you in the eye from your contaminated projects. Only through experience, following the growth parameters and sterility considerations outlined here and keeping a sense of play, will you find what works best for you.

If you are on a time crunch my suggestion is to go over this whole section several times to get a thorough understanding of exactly what is needed at each stage of work. Perhaps create a flow chart or time line of things to expect down the road so that you have everything you need for each step of the way. Once you have all your resources gathered, attempt several of the techniques listed here at the same time. This may seem daunting, but if you are thoroughly prepared and have thought the process through, you will likely have success with at least one of the techniques. Further, taking this shotgun approach will quickly help you learn from your successes as well as from your mistakes. Take good notes of all you do, be as clean as you reasonably can and if worse comes to worst ask for help from the kind folks on one of the online forums listed at the end of this article.

## CARDBOARD SPAWN

You can attempt this cheap and simple technique with any species of saprophytic fungus. As it turns out, the glue in corrugated cardboard is wood based, and is thus a viable food source to decomposer mushrooms (though surprisingly not much else). What this means is if you follow the simple steps below you will get a bit of mycelium in not a lot of time, for not a lot of dough and with limited risk of contamination. Great! This technique is best used with fresh mushroom pieces (preferably the stem bottoms, though any part will potentially work) though it can work with plugs as well. Done right, this technique is quite successful and easy.

1. Obtain a corrugated cardboard box from the US or Canada (cardboard from other countries may contain nasty chemicals and inks you don't want). Try to get the most plain and ink-free box you can.
2. Remove all tape and stickers, and flatten the box.
3. Soak the cardboard in hot water until it is thoroughly saturated (roughly 60 minutes).
4. Remove from water and let the excess water drip off.
5. Remove one side of the cardboard, exposing the corrugations.
6. Evenly disperse your fungus throughout the cardboard (say, one piece every three inches).
7. On top of this fungus layer, now place another layer of saturated corrugation and backing so that the fungus is sandwiched between two layers of corrugation.
8. Store the sandwich in a dark, warmish place where it will get air exchange, retain its humidity and not dry out. One example would be in a Tupperware® container kept above the refrigerator (this space is often warm due to heat rising from the refrigerator's motor). Be sure to circulate the air in the container at least once daily and to spray with water as needed to keep moist.
9. In several days or weeks the mycelium should start to grow out onto the cardboard, appearing as distinct or fluffy white lines.

Once the mycelium is actively growing, it should soon be moved onto another food source. In my experience, this technique works for

Fig. 6.1:
Hypsizygus
ulmarius
*grows rapidly*
*on saturated*
*cardboard.*
CREDIT: PETER MCCOY

a short period of time before the mycelium begins to stagnate. Adding a sprinkling of spent coffee grounds to the cardboard can help provide an additional nutrient boost and support mycelial growth. However I would recommend using cardboard spawn as a short intermediary technique before quickly moving on to sawdust or bunker spawn (described below). Some people might also place this cardboard between layers of fresh hardwood chips in a cardboard box, letting it sit undisturbed for several months. While this is an option, it is a bit slow. See what works for you.

## COFFEE OR SAWDUST SPAWN

Many different substrates have and can be used to grow mushrooms. These include tea, newspaper, old phone books, egg cartons, fabrics, hemp, twine, corncobs, grass clippings, hair, leaves, tobacco stalks, brewery waste, manure, coconut coir and many other types of organic material. While this list makes the options rather wide, in this section we will focus on two bulking substrates that are both reliable and accessible: spent coffee grounds and hardwood sawdust. If you have enough source mycelium to play around with, try following the basic procedure listed below with some of these alternative substrates and see what works for you. Different species grow better or faster on certain substrates, but even this can vary depending on the strain, so in some ways you never know what will work until you try it.

The standard bulking spawn material in the professional world of cultivation is hardwood sawdust, often supplemented with wheat or rice bran to provide additional nutrients. Alder and maple are the most commonly used types of wood. Not only are these tree types more easily acquired as a commercial product, but also they are softer types of hardwood which fungi can break down faster than harder species like oak. Many species of cultivated fungi can use these two types of wood with a minority of species also being cultivatable on softer, coniferous wood. Other hardwoods can also be used, depending on the species. Such specifics are best referenced in the cultivation books listed in the recommended reading at the end of this article. Alder sawdust may be available cheaply from a landscaping supply company, wood mill or furniture manufacturer. Some say the fresher the wood is the better in order to avoid the risk of contamination from other fungi in the air. Others suggest waiting a few weeks from the day the tree was felled so that any antifungal properties naturally found in the wood will degrade. I've known people to use very fresh and very old wood with mixed results. The standard answer then is to use wood that is between two weeks and two months old. But shop around and work with what you've got.

Another easily accessible substrate that many amateurs use is freshly brewed coffee grounds, often available for free from your local coffee house. Essentially pre-pasteurized during the brewing process, these grains can be used quite readily with many oyster species. I have also gotten turkey tails to grow on coffee grounds as well and imagine other strong saprophytes might take to this medium. Ideally the grounds would have been freshly brewed, though not hot when inoculated.

The steps for bulking out your spawn using any of these substrates are the same regardless of which are used and follow this basic process:

1. Wash your hands. It doesn't hurt and only helps. Follow other cleanliness guidelines as desired.
2. Moisten your substrate. The ideal moisture range for fungus to thrive is a 60–65% saturation level of the substrate. Coffee grounds will be as wet as you get them from the shop, though try to find some that are moist but not dripping more than a drop or two when squeezed. For sawdust, you will need to spray down the material until it too is wet enough to hold its form but breaks easily apart. The setup I use for this spraying process is a basic window screen

supported on two blocks. On this screen I pile a layer of sawdust two to three inches thick and mist it with a garden hose, stopping often to mix the water in and check the moisture level, then waiting 20–60 minutes for the water to be absorbed. In time, you will quickly learn to feel when the substrate is sufficiently saturated. You can also check your moisture content by weighing out about 100 grams (4 ounces) of wet material, drying it thoroughly in an oven, and then measuring the final weight. If the end weight was, for example, 35 grams (1¼ ounces), then the original wet material was at 65% saturation.

3. Optional pasteurization. In professional cultivation guides you will be instructed to pasteurize the substrate at this point to kill off competitor bacteria and fungi that are present there. In my experience with the *Pleurotus* species, *Hypsizygus ulmarius*, *Stropharia rugosoannulata* and several other strong growers, this is not necessary as long as a fair amount of source spawn is introduced. Roughly one part source spawn to every five parts substrate is a good starting ratio. Larger concentrations will result in faster colonization and a decreased risk of contamination. I have repeatedly had results with low contamination using just plain, untreated, moist sawdust. My belief is that if the fungus can survive these contaminants in nature, why not here? Of course, you will need to experiment yourself. If this process is not working for you, you may opt to look into other options or pasteurization techniques — or save up to buy spawn kits. To pasteurize your sawdust, place the pre-moistened sawdust in a canning jar or heat-tolerant plastic bag (available from mushroom growing suppliers), place the jars into a pot filled with cold water. Heat the water bath to boiling and insert a meat thermometer into the center of one of the sawdust jars. Once the temperature reaches 120°F, turn off the heat under the water bath. The temperature will continue to climb but should not go over 175°F. Maintain an internal temperature in the jars of 140–160°F for one hour, turning on the heat under the water bath if the temperature drops too low. After an hour the sawdust should be pasteurized and, once cooled, ready to be inoculated. Coffee grounds, if fresh from the shop, should essentially be pre-pasteurized for you.

4. Pack it up. Now you will want to put the substrate in a suitable container. Pretty much anything will work as long as it is clean,

will maintain the right moisture level and provide for some means of air exchange. Mason jars with special lids or high-heat-tolerant plastic bags with built in filter patches are the professional route. I prefer a simpler approach. One option is to stuff a clean one-gallon Ziploc® bag (the kind with the slider closer, not the old snapping kind) about three quarters full of the wet substrate. When I go to slide it closed I stuff a tuft of synthetic Poly-fil® (quilt batting) about the size of a closed pinky finger into the end of the slider to act as my filter. This setup works quite well for me. One could also cut a hole in a plastic bag and tape a square of Tyvek® (available as shipping envelopes at office supply stores) over it to provide for air exchange. Other container options include trash bags, plastic tubs or buckets with holes drilled for air exchange or mason jars with Poly-fil® stuffed through a hole in the lid for air exchange.

5. Add some of your source mycelium. This is where all your original scrounging comes in to play. You should play around with using the different types of source mycelium. Starting with a commercial kit is a great place to begin bulking, but I have even had success just throwing chunks of torn up oyster mushrooms into a bucket of coffee grounds and covering them with more grounds. One key here is that the more mycelium you introduce the faster the substrate will be colonized and the smaller your risk of contamination will be. Once introduced, shake up the contents to evenly disperse the mycelium throughout the substrate. Alternately, introduce your source spawn, mix, then pack into your container.

6. Place in a suitable environment. Different species have different temperature and humidity requirements for optimal growth. While these should be taken into consideration, unless you have a dedicated growing space, a warmish, not too dry place should be fine. You will also want to place the fungus out of direct sunlight.

7. Wait. This is the hardest part. In the coming weeks you will want to check the bags often. In the ideal situation you will begin to see white threads of mycelium radiating out from each patch of source mycelium. These patches will expand over several weeks to encompass the whole bag. If they are growing out healthily, it is a good idea to shake the bag up at least once through the process (at around 30 percent colonization) to help distribute the mycelium even more

and achieve faster colonization. Things to look out for are any signs of discoloration from competitor molds (e.g. blue, green or gray patches) and slow or no growth. Also, if you wait too long, mushrooms will begin to form in the bag. This is a sign that the fungus has run out of food and needs to be moved onto the final substrate.

Fig. 6. 2: Stropharia rugosoannulata *ripping through untreated plain sawdust.* CREDIT: PETER McCOY

Fig. 6.3: *Improvised filter in Ziploc® bag.* CREDIT: PETER McCOY

Fig. 6.4: Pleurotus pulmonarius *growing on spent coffee grounds.* CREDIT: PETER McCOY

If things don't work out at this stage and all of your bags become contaminated you might have to start all over. This would be a good time to review your notes, trying to determine where things went wrong so that you do not repeat the same mistake. See the trouble-shooting section at the end of this article for ideas on improvements. However, if you are successful you will likely have bulked up your my-celium to such a critical mass that your success in the next step is much more assured. At this point, you will now have a bag full of happy, and hungry, mycelium ready to be moved ASAP on to the next, and final, stage.

### Step 3: The Final Substrate

Now that you have bulked up your mycelium and actually have some-thing substantial to work with, you can move on to the final stage: giving the mushroom a longer-term source of food. With this last stage the fungus will now have something on which it can feed for months or years while it starts taking care of your pollution problem.

Whether you are trying to clean up a polluted waterway or dirty soil, one of the most effective and versatile options is to create *bunker spawn*. This term simply means burlap sacks filled with a combination of bulk spawn, wood chips, cardboard and pasteurized straw.

As with sawdust spawn, the age and species of wood chips you use should be appropriate for the fungus you are working with and will ideal-ly have been presoaked for 24 hours to increase moisture content. The straw you use will need to be treated against contamination, as straw is rather susceptible to attack from competitors. Traditionally, the straw was submersed for an extended time in large vats of heated water to pasteurize the material. However, in recent years, a process known as *cold water pasteurization* has been developed to cut out needless waste of fuel and make cultivation one step easier for the average cultivator.

#### COLD WATER PASTEURIZATION

Simple and scalable, this method is a breakthrough low-tech cultivation technique. Essentially a fermentation process, cold water pasteuriza-tion kills off competitors to your spawn by simply  submersing the straw in water over a period of days. During the submersion process anaerobic bacteria present on the straw thrive by eating all the aerobic (oxygen-loving) bacteria. When the water is removed after a week, the anaerobic bacteria die, leaving clean straw to use for inoculation.

1. Line a garbage can (or any hard, upright container) with a heavy-duty (three mil) trash bag.
2. Wet the inside of the bag down with a bit of water. Ideally, this would all be non-chlorinated water or, better yet, well water. Alternatively, if you leave tap water out for a day, the chlorine will evaporate.
3. Simply place the straw in another dry trash can, insert a weed whacker and chop the straw into one- to three-inch pieces. Using this machine increases the straw's surface area and enables better colonization and easier handling later on.
4. Place the dry, chopped straw inside the trash bag inside the can and fill the bag with water, covering the straw.
5. Add a clean weight on top of the straw to keep it submerged.
6. Put a lid on the can and keep it in a warm place, ideally in the sun.
7. Wait seven days.
8. At this point the water should be discolored and stinky. This is good. You will now want to turn the can upside down and drain the water off. Once empty, twist up the top of the trash bag and turn the can upside down to allow the remaining water to drip off for two additional days.
9. At this point the straw is ready to be used directly as a substrate or in a mix of bunker spawn (see below). If using alone, follow the same basic procedure as with bunker spawn above.

Fig. 6.5: Pleurotus pulmonarius
*growing on cold pasteurized straw,*
*inoculated with sawdust spawn.*
Credit: Peter McCoy

## BUNKER SPAWN

Now that you have treated your straw, you are ready to create bunker spawn. In its simplest and quickest form, you would only mix the straw and bulk spawn together and stuff it in a burlap sack (also conveniently procured from your local coffee roaster). The steps below, however, describe a more thorough process that provides the mushroom with a longer-term food source.

1. Soak pieces of clean, ink- and tape-free corrugated cardboard in hot water until saturated. Let sit for now.
2. Wash your hands and follow any cleanliness procedures desired.
3. Lay out a large tarp and clean it off. This could be as simple as spraying it down, though the more sterility conscious will wipe it down with 80% isopropyl alcohol for added cleanliness.
4. Spread your treated straw and soaked wood chips together on the tarp. Depending on your needs you might add more straw than chips, or vice versa. The former provides faster colonization, while the latter offers longer viability.
5. Break up the bags of bulk spawn and dump most of their contents over the straw and chips. Observe a roughly one-to-five suggested inoculation ratio, but work with what you have. Reserve a small portion for step #7.
6. Mix all of this together thoroughly by rolling it all back and forth within the tarp.
7. Take the wet cardboard from step #1 and strip one side, exposing the corrugations. Introduce a small chunk of bulk spawn to these corrugations and roll up the cardboard like a burrito. Make a good number of these.
8. Stuff the mixture from the tarp into your burlap sacks. As you do so, introduce two to three spawn-filled burritos into each sack. The burritos will serve as mycelia hubs in the bags and will facilitate better growth within them.
9. Once full, stitch the bags closed with heavy-duty thread.
10. Pile these bags on the ground or on a pallet outside in the shade (the warmer the area the better).
11. Wait.

Fig. 6.6:
Stropharia rugosoannulata *bulk spawn growing through wood chips and burlap.*
CREDIT: PETER McCOY

From here the mycelium will hopefully do rather well. Like sparks from a fire, the mycelium will quickly jump from the cardboard to the straw to the wood chips until, in time, the entire bag will be consumed by mycelium as the fungus digests the organic material. Once the bags are well colonized, their contents (that you have worked so long and hard to create) can now be used in a number of ways for remediation.

## DIY Fungal Remediation

You've come this far, the goal is in sight, now is the time to put your fungus to work! As described in the beginning of this article, you can use fungi to clean up polluted areas in numerous ways. An easy way to begin is through the direct application of your newly created bunker spawn. The techniques below are general guidelines and, while seemingly short, contain numerous possibilities for creative interpretation. Fungal remediation techniques are not set in stone; they are constantly being improved upon by people just like you. Play around with these techniques, and adapt to your circumstances as needed.

## Fungal Water Filtration

If you have a polluted stream or pond, bags of bunker spawn may be submerged in the waterway to act as a living filter, thereby trapping chemical and biological pollutants as water passes through them. For flowing water, you will want to form a solid barrier that forces most of the water to channel into the bags instead of around or over them. In standing ponds, colonized straw bales or bunker spawn can be introduced to potentially trap pollutants, including heavy metals.

Similarly, furrows can be dug in the land and filled with mycelium to act as a standing filter that will treat accumulated surface water before it soaks into the soil. Furrows can take the form of long trenches that follow the contours of the land, of terraces upon a hillside or as a single large depression/shallow pond. These furrows (or swales as they are called in permacultural practices) can be filled with mycelium by layering bunker spawn bags or by forming a mushroom bed. A *mushroom bed* is simply the layering of plain cardboard; two to three inches of suitable hardwood chips; an even layer of bulk or bunker spawn followed by two to three more inches of wood chips. A second layer of spawn and a third layer of chips can additionally be added if desired. Finally, the mushroom bed should be thoroughly watered and covered with six inches of straw to prevent drying out. This bed will last for years as long as fresh chips and water are introduced as needed.

## Fungal Soil Remediation

If you have a problem with polluted soil, this dirty earth could be dug up, piled and then layered with the bunker spawn to begin treatment. By forming a lasagna with the spawn and contaminated soil you may be able to provide a suitable habitat for remediation. This method has sometimes proven successful but may require experimentation to enhance results. As noted when I described the concept of "training" with spore slurries, the introduction of clean soil, pasteurized straw or other organic material might also help the mycelium survive better in heavily contaminated soil.

Alternately, bags of bunker spawn could be expanded in one more generation to more bags, this time with the addition of the polluted soil as well as more treated straw and wood chips. This is where the sense of play must be emphasized; experimentation with different ratios of contaminant to fresh substrate as well as fungal species or strains may be necessary to find what is most effective. The Amazon Mycorenewal Project performed experiments of this kind and determined that a one-to-four ratio of polluted soil to clean substrate yielded the best results in their work with, using oyster species to remediate oil-saturated jungle soil. Ultimately, however, I do encourage you to research what species and techniques are effective against the pollutants you face. While experimentation is critical for the advancement of understanding, there is a wealth of research that has been done thus far to save you the time of repeating experiments that are known to fail. Please refer to the recommended reading below as a starting point for future research.

## Erosion Control

In areas where erosion may present a problem, laying a mushroom bed may help to stabilize soil structure. As the bed will need to survive for years in order to be most effective, native species that are adapted to the environment and climate of the area should be selected. As opposed to the beds used for remediation, beds laid for erosion control might benefit from direct contact with the ground (i.e. no cardboard bottom) and from the introduction of native grasses on top of the straw layer to further enhance soil structure.

## Where Do We Go From Here?

Because fungi are so foreign to the minds of the uninitiated in most of the western world, learning to cultivate them can become a truly herculean effort. This is perhaps the biggest hurdle to the advancement of the science of mycology, let alone the praxis of Radical Mycology. We must accept that we will make mistakes, that spawn will get contaminated, that our experiment might not work or the remediation might not be successful. Once we make this leap of faith, and begin working with the fungi, however, the limits before us are truly unknown.

Sadly, this is not the message most people receive willingly when approaching mycology. While I have seen a shift in this sense of appreciation for the fungal kingdom in recent years, we still need to go a long way before more of the world has access to these resources and skills. Hopefully at this point in this article you have come to realize how simple and amazing fungi can be to work with once some basic concepts are grasped. And hopefully you begin to put these ideas to work and try them out for yourself.

Mycology is one of the youngest natural sciences and, as such, discoveries are made within it all the time. There is so much yet to be understood about the remediative properties of fungi that the world cannot simply wait for a small number of professional researchers to figure it out — and then patent the information. Without concerted (while somewhat playful) experimentation and research by people just like you and me, progress will only continue at a snail's pace. Thus, I call you to join the Radical Mycology movement and empower yourself to become "be-mushroomed."

It would be amazing in the coming years to see remediative applications of mycology blossom into common knowledge around the world. An invaluable resource for communities, mycology could certainly help us out of many of the binds in which we currently find ourselves. In my ideal world, every major city (at least) would have a community lab space and fungal remediation team ready to go at a moment's notice when the next oil spill strikes. Such infrastructure will require more people informed about these practices as well as having community access to resources as they are needed. In order to make that a reality, we must no longer look to the pedagogues of mycology for assistance but toward each other for our collective understanding of the subject. And we must look toward the fungi themselves as they are there for all of us and will work with us if we work with them. The future of fungal remediaton is uncertain, but by working with these skills and spreading this information we all will surely take another great step toward a better world.

In closing, I'd like to present a rough outline of considerations to take when approaching the use of fungi for remediative purposes. Take these items as starting points for further investigation and your own studies.

1. Read this chapter thoroughly and begin looking to other resources. While the information given here (once it is thoroughly understood) is enough to get one going, some people will feel more comfortable taking on a new skill if they have an even greater

depth of knowledge. It is up to each reader to decide. Mycology is an infinitely fascinating and mysterious field of discovery, and the possibilities for working with fungi extend far beyond the options I have presented. The sources and recommended reading at the end of this article are invaluable resources for gaining a more solid foundation in fungal cultivations and remediation. The online community, especially, is a wealth of information for people seeking low-tech and simple cultivation techniques. Some of these sites cater to the psychedelic growing community, but information and techniques presented on them readily translate to cultivation for remediation purposes. Once you become a member (for free) you will be surprised by how helpful people can be in answering questions and helping overcome problems.

2.  Determine what exactly you are up against. Have water, soil or harvested mushrooms tested in a professional lab. This will not only tell you what you are working with but will also set a baseline to see how effective your end results are. Without this step your selection of fungal species will be a shot in the dark, and the actual results of your remediation efforts will be unknown. The Agency for Toxic Substances and Disease Registry at the US Center for Disease Control can test a thoroughly dried mushroom for its heavy metal content for US$35–150.[7]

3.  Determine what approach, species and strain you will use. What have other people done to deal with a pollution problem similar to yours? What species are most effective at dealing with your contaminant? Is there a native species or strain you can procure that is adapted to the your own environment or substrate? Might you use multiple fungal species to counteract several different toxins? How will you keep the fungi wet and out of direct sun? Look up pictures and designs of other remediation projects, and get ideas to make your project as effective as possible.

4.  Determine how you will track your progress. How will you know you are successful? Are there multiple ways to approach the problem? What kind of experimental design can you come up with to compare results between different setups and to account for variables? What follow-up tests will you perform to measure reduction in pollutants?

5. Acquire the needed tools and infrastructure for your level of myco-logical work. The specific considerations in designing a small DIY community lab will vary depending on needs, funding and time commitment. The differences between catering to a one-time re-mediation project or creating a continual stock of fungi (good for emergency scenarios) are quite vast. Talk to your crew and decide how invested you are in furthering this work in your collective future. You may also consider getting the appropriate level of hazardous materials training and equipment to stay safe in your projects.

6. Cultivate fungi! If you are in contact with a mushroom farm that can supply the species you need, you might be able to end there. However, chances are that for any appreciably sized project you should consider cultivating your own fungi to cut costs. Following the techniques described above will hopefully get you where you need to be. A big consideration, however, is whether you will want to maintain a constant supply of fungi. There are many benefits to this, especially when unexpected disasters strike. Yet it can become a bit like taking care of pets, as you will be quickly growing so much mycelium that you may overrun your home with bunker spawn without a planned growing schedule. Based on experience begin to imagine how long it will take to grow a considerable amount of mycelium from a small amount of source spawn. Plan accordingly based on space, time and need.

7. Remediate and Teach! This is perhaps the most important step. I say this not only because you are finally applying months of dedi-cated, and at times challenging, attention to detail to help heal the land around you — but because you can now share what you have learned with your community. Now that you have passed through the fire unscathed you can live to tell others of the powers of fungi and help spread their spores a little farther. Mainstream society isn't helping the fungi, and the professionals don't know the people you know, so it is up to you to help teach and inspire others to learn how to cultivate themselves. Or at least let them borrow *Earth Repair*.

I hope this article has inspired you to explore the fascinating world of mycology and perhaps learn more. For the more one works with fungi, the more they will teach you and the easier the road becomes. I wish you the best of luck in you efforts; may your spores spread wide and strike hard.

## Supplies Checklist

### Spore Slurry

- ☑ bucket or jug
- ☑ mushrooms
- ☑ molasses
- ☑ salt

### Bulk Spawn

- ☑ alder sawdust (from landscaping supplier, furniture factory, wood mill)
- ☑ cardboard (ink- and tape-free)
- ☑ coffee grounds (from coffee roaster)
- ☑ hose, screen and supports (for wetting sawdust)
- ☑ laundry baskets, buckets
- ☑ Poly-fil® (from sewing or quilting store)
- ☑ source spawn
- ☑ Ziploc® bags, mason jars or buckets

### Bunker Spawn

- ☑ bulk spawn
- ☑ burlap sacks (from a coffee roaster)
- ☑ cardboard (ink- and tape-free)
- ☑ needle and thread
- ☑ straw bale
- ☑ tarp
- ☑ trash can with lid
- ☑ weed whacker (to chop up straw, optional)
- ☑ wood chips (use a suitable species, no bark)

## Short Species List

### SHAGGY MANE (*COPRINUS COMATUS*)

Mycelium grown on sawdust can easily be buried in the ground to establish patches of this species. Favors the edges of grassy areas, polluted soils and eroded and disturbed areas. Known for its ability to break through asphalt (spore slurries anyone?). Useful for metal-polluted soils to sequester arsenic, cadmium and mercury. Studies have also shown it to inhibit *Aspergillus niger*, *Bacillus species*, *Candida albicans*, *Escherichia coli*, *Pseudomonas aeruginosa* and *Staphylococcus aureus*.

Shaggy mane spawn grows best at 70–80°F.

### ELM OYSTER (*HYPSIZYGUS ULMARIUS*)

Produces strong cellulases that Stamets suggests may be useful for paper products, dioxins and wood preservatives such as chromated copper arsenate (CCA). Incredibly easy to cultivate on many substrates.

Elm oyster spawn grows best at 70–80°F.

### SHIITAKE (*LENTINULA EDODES*)

While not as easy to grow using the techniques above, if spent kits can be acquired they can be packed into burlap sacks to act as water filters. Shiitakes are known to break down polycyclic aromatic hydrocarbons (PAHs), polychlorinated biphenyls (PCBs) and pentachlorophenols (PCPs).

### KING OYSTER (*PLEUROTUS ERYNGII*)

Known to breakdown a variety of toxins (including the base agent in Agent Orange, 2,4-dichorophenol), it is recommended for soil remediation and filtration applications.

### PEARL OYSTER (*PLEUROTUS OSTREATUS*)

The iconic remediator mushroom, this species is a prolific spore producer and aggressive decomposer of a range of pollutants, notably PCBs and the PAHs found in petroleum products, pesticides and many other toxins. It is also known to sequester cadmium and large amounts of mercury. This is one of the easiest mushrooms to cultivate and easily suited to the techniques described above.

### PHOENIX OYSTER (*PLEUROTUS PULMONARIUS*)

This species has been shown effective against dioxins and TNT and is known to sequester cadmium, mercury and copper. Like the other oysters, it is very easy to cultivate.

All three oyster species grow best at around 75°F.

## Turkey Tails (*Trametes versicolor*)

Known to sequester mercury and filter *Escherichia coli, Listeria monocytogenes, Candida albicans* and *Aspergillus* species. Has also been shown to effectively break down many PAHs including pyrenes, fluorine and styrene as well as pentochlorophenols, TNT, CCA, dioxins, anthracenes and persistent organophosphates. An easy and aggressive species to cultivate.

Turkey tail spawn grows best at 75–85°F.

### Sources and Recommended Reading

#### Sterile Cultivation

Paul Stamets. *Growing Gourmet and Medicinal Mushrooms*, 3rd ed. Ten Speed, 2000.

#### Fungal Remediation

Harbhajan Singh. *Mycoremediation: Fungal Bioremediation*. Wiley-Interscience, 2006.

G.M. Gadd, ed. *Fungi in Bioremediation* (British Mycological Society Symposia). Cambridge, 2008.

Paul Stamets. *Mycelium Running: How Mushrooms Can Help Save the World*. Ten Speed, 2005.

### Websites and Online Forums

+ Radicalmycology.com
+ Fungiforum.com
+ Mycotopia.net
+ Shroomery.org
+ Permies.com/forums/f-39/fungi

*Peter McCoy is an artist, musician, street medic and rebel mycologist who has been studying the power of the fungal kingdom off and on for over a decade. What started out as a personal hobby has developed in recent years into an all-out campaign to bring simple fungal cultivation for self-sufficiency and remediation to the masses. Peter is one of the co-founders of the Radical Mycology meme and can be found teaching workshops throughout the US on DIY mycology. He is currently based out of the Cascadia bioregion and loves it there.*

## Focusing on Soil Remediation and Water Filtration with Fungi

BY JA SCHINDLER, FOUNDER OF FUNGI FOR THE PEOPLE

Partnering with fungi can open the door to an accelerated process of soil and water healing. The natural behavior of some fungi to work within and support a larger community of organisms seems to be the key part of their role. The lignin-decomposing, white rot *Basidiomycotina*, fungi with known abilities to break down complex plant cell structures, have been the primary case studies thus far in mycoremediation, but a much wider array of soilborne fungi and endophytes with lesser understood natural roles are becoming more important subjects of inquiry.

Those fungi in the group of plant lignin-decomposing *Basidio-mycotina* have been employed mostly for two reasons: first, because this is the group to which most cultivated mushrooms belong, so there is a great deal of information available on how to grow them. The second reason is that some of the more aggressive plant lignin-decomposing fungi (oysters, turkey tail, some *Stropharia*) already have highly developed mechanisms and digestive enzymes for pulling apart and degrading very complex carbon based compounds, as well as trapping and digesting an array of bacteria. The lignin found in dense trees turns out to be very similar in structure to many of the heavy toxins we need to deal with in soil and water remediation, so these fungi can lend themselves readily to the task. Plants and fungi have coevolved over millions of years so that we live immersed in a rich world of plant and microbe diversity. This diversity is lost in most environmental degradation, and the forests and wild healthy places which remain harbor the greatest gene pools. There are an enormous amount of fungi and other microbes that have never been studied nor identified; the job is immense, and the rapid destruction of intact ecosystems is making the work even more important. The diversity of industrial pollutants is also enormous, and their steady contamination of clean water and soil has only increased in the past century. This wild place of toxicity that we find ourselves in has inspired serious interest in the roles and abilities of fungi to regulate. Fungi not only like to disassemble; they can be very social creatures, living in direct contact with thousands of other very alive beings. We are finding excellent examples of species degrading specific toxins, as well as many species degrading a range of toxins. The search has been on for a while to marry species and technique with appropriate situations. Working with native fungi in a way that encourages a healthy diversity of other microbes seems to make the most sense in remediation practices.

One of the clear roles that fungi played in early mycoremediation studies is supporting and energizing other organisms into remediation activity, just as they often seem to support and energize those same organisms into living healthily in native ecosystems. It is as though the enzymes secreted by many fungi actually stimulate aggressive toxin degradation by other microbes present in the soil as well as plants. If there is a sensible order to it, research findings support the outlook that some fungi can degrade complex toxins such as 4–5 ring PAHs into simpler toxins more bioavailable to other organisms. Starting with a supporting role, fungi can trigger an aggressive chain of commands in the soil ecosystem toward soil health.

## Approaches

There have been many approaches proposed and carried out for working with fungi as a natural remediator in the past few decades. Various areas of focus have been considered such as degrading polycyclic aromatic hydrocarbons (PAHs), polychlorinated biphenyls (PCBs), polycyclic polymers (PCPs), TNT, neurotoxins, synthetic dyes, waterborne and airborne pollutants. A range of research projects are cited at the end of this article.

Most soils and waterways with the need and potential to be remediated are sites where toxic industrial activities, oil spills or other accidental discharges of toxic material have left an accumulation of heavy carcinogenic compounds. In many cases there may be plants and fungi actively growing in the contaminated soil, and in other cases there may be heavy deposits of toxins visible to the naked eye. At either end of the spectrum the first step is to get an idea of the level of toxicity and the pH of the soil. If it is affordable to have soil tests taken professionally through a private environmental firm it is worth developing an idea of what you are working with so that later tests can give a better idea of the effects.

Other soil testing resources are local colleges and universities, and many cities have a protocol for safe toxicity levels. If funding is tight, other options can include buddying up with soil research students looking for projects or getting to know plant and bacterial signatures in relation to toxicity. If you have access to none of these options and know that the soil is toxic, it is at least worth getting your hands on a soil pH test kit from a garden supply store.

The term pH is an expression of nutrient availability through the reading of available hydrogen in the soil; pH reading is a direct answer to the diversity of organisms either currently living or potentially living in that soil. The closer to a pH of 7.0 that you can get, the more biologically active soil will naturally become. One of the best ways to adjust soil pH is to amend with organic material such as straw, leaf mold, manure or wood chips from broadleaf trees such as alder, maple and oak. Layering these materials on top of the soil and allowing rain, microbes and insects to slowly incorporate them is very effective. Turning or tilling these materials into the soil for a more homogenous mixture is even more effective at activating the biological process. Additions of organic material at a rate of at least 20 percent and up to 80 percent of the pile or site will ignite an array of biological activity.

Most studies have shown that amending with organic matter is a very important factor in activating toxin degradation with or without introduction of mycelia. This addition of material is referred to as *bioaugmentation*. Consider that you will be contaminating whatever healthy material you add to the site. I would recommend addressing soil pH before amassing a load of fungal mycelia.

Avoiding temperatures above 120°F and below 32°F will allow for a much broader range of biological activity. Temperatures above 120°F will kill off many microbes and promote only heat-tolerant (thermophilic) fungi and bacteria, which is a much smaller range of organisms. Dropping below 32°F will bring on dormancy, and the pile or site will do little until temperatures rise again.

To accomplish good temperature range just think about composting and garden mulching. With composting, fresh new piles of material heat up very quickly at the core, but can be discouraged from going too hot by dismantling the pile and allowing cool air to enter while hot air escapes. Slow composting is preferred for toxin decomposition, and can be accomplished by keeping piles to under 18 inches and not actively turning them. To keep temperatures up in cold months think about how you might protect plants and garden soil over the winter. By piling layers of straw, covering with cardboard or building a row cover, you can keep frost from affecting the bioactivity of the site. Available oxygen is also a huge factor, and steps may need to be taken to aerate the pile or site. Ventilation tubes of thick cardboard can be used in a pile or buried in a deeper site. They are often available as waste from carpet stores and will degrade naturally as the site matures.

Previously in this chapter, Peter McCoy gave a briefing on a range of approaches that might make sense in applying fungi to your remediation project, so I will describe a few projects I am currently working on, as well as a pollution source filter. As Peter described, there are techniques involved in growing out mushroom mycelium for use in remediation, but there are some low-tech alternative approaches more recently uncovered that are helping tremendously with this issue. One of the big blockades to widespread mycoremediation is the lack of individuals skilled in fungal culturing, which is why Peter and I commit a lot of energy and time in sharing the knowledge. Since 2011, I have been hosting workshops on cultivation and remediation and focus heavily on low-tech, low-cost approaches.

Ethically there are a few factors to consider when working with aggressive fungi. First, working with fungi local and native to the contamination site not only ensures that you won't be introducing an invasive species, it also allows the possibility of working with a fungus that is adapted to the the toxins present. Second, there are research results showing that fungi that are parasitic to vascular plants may be strong remediators; the inherent problem is that promoting their soilborne reach could lead to the demise of a whole-system approach, potentially worsening the situation. There are more techniques being tried all the time, some of which involve working with fungi to degrade toxins at the source before they enter the soil or water. These seem to be better approaches for applying foreign or parasitic fungi.

Human history is filled with stories and recipes for partnering with plants and fungi for natural human health, and recent years have seen these traditions reborn in Western culture. There is much to learn from our ancestors, and in turn, much to discover in how to partner with Earth's beautiful other beings in restoring and maintaining eco-system health.

## Project: H.A.R.P. Toxic Compost Cleanup Location: Portland, Oregon, Project Start Date: June 1st 2012

### SCENARIO

During the 12th annual Village Building Convergence in Portland, Oregon, I hosted a mycoremediation workshop at the Healing Arts Resource Project (pdxharp.org). H.A.R.P. is a project where a collective of natural health practitioners come together to share in their healing arts. They had just purchased a new home for the project in the Sellwood neighborhood, the former site of a conventional hair salon. One of the big problems with the new location was a very large and unhealthy compost pile in the garden area of the backyard that included tar-based roofing shingles, plastic grocery bags, credit cards and human hair from the former hair salon.

Earlier in the week another presenter from Chicago, Nance Klehm, had deconstructed the old compost pile and moved the soil aside, erecting a new pile with healthy ingredients and a plan to keep it that way. I attended her workshop and was impressed with her knowledge and outlook on soil health, so I was left with wanting to remediate the pile of soil and the site where the composting had taken place. Two approaches were carried out: a top-down mushroom garden bed on the site of the old/new compost piles, and an in-vessel remediation of the old compost.

Twenty soil samples were taken from different points around the compost site and in the old compost, three of which were from directly under where the mushroom garden bed was to be installed. It should be noted that no plants were growing in any of the soil up to two feet from the compost site, and no plants in the pile itself, an ominous phenomenon in Portland's lush climate.

The samples were taken at a depth of one to three inches, and two samples were taken at a depth of six inches. Half of each sample has been sent to a soil lab for testing. The soil tests have not been returned from the lab yet, but I am using the other half of each sample to assess for present fungi in the soil willing to be cultured. Testing at home showed the soil pH to average 7.2. At three-month intervals for a year

I will take samples to be tested from the same sites, and that information will be published on my website (FungiForThePeople.org).

### PILE SITE: *STROPHARIA RUGOSOANNULATA* (SRA) TOP-DOWN MUSHROOM BED

Once the samples were taken we installed a two-by-eight-foot patch of SRA on a substrate of Douglas fir wood chips along the backside of the compost site between the structure and the brick wall that edges the property. It will be completely out the direct sunlight, and the brick wall will insulate, heat and help maintain moisture through the dry summer months.

Creating the top-down mushroom bed was fairly easy:

**1st Step:** We laid down a ¼-inch sheet of corrugated cardboard as a base for the shape of the bed. This will give the culture of SRA a chance to dominate the fresh Douglas fir chips before other fungi move up from the soil.

**2nd Step:** We laid down a three-inch layer of pretty freshly cut Douglas fir wood chips on top of the cardboard and watered it for about ten minutes.

**3rd Step:** Two SRA grow bags of 1.5 gallons each were spread out evenly across the layer of wood chips. The grow bags consisted of a matured mycelial mass of SRA grown out on a pasteurized mix of alder sawdust and Douglas fir wood chips, so the fungus was already acclimated to the wood chips on a non-sterile substrate.

**4th step:** We then covered the bed with another two inches of fresh Douglas fir wood chips and watered the patch in for about 15 minutes before covering with another layer of cardboard to keep the moisture in.

The technique is the same as you would use if you were starting a mushroom garden bed anywhere, only with the intent of the mycelia eventually diving into the soil layer below and activating the bioremediation process. SRA does very well in non-sterile conditions, seemingly stimulated by other organisms, and in turn supports a healthy diversity in the soil ecosystem, a vital process in complete decomposition of most toxins.

Figs. 6.7 and 6.8: *Laying out wood chips and inoculating with king stropharia (*Stropharia rugosoannulata*)*
CREDIT: JA SCHINDLER

## DEALING WITH THE OLD TOXIC COMPOST

The second setup we worked out was to use *Hypsizygus ulmarius* as a remediation kick-starter for the old pile of toxic compost. The approach we used was to fill two 25-gallon half barrels with the old soil sandwiched between layers of *H. ulmarius* spawn. One of the barrels

was filled directly from the pile, and the other was screened with a
½-inch screen to remove bulk contaminant particles, such as combs
and larger pieces of tar shingle.

The half barrels have a two-inch drain hole in the bottom, so we
propped them five inches off the ground and placed a sediment catch
under both of them. To simulate natural conditions as much as possible, the barrels were left outside in the open with just a layer of
corrugated cardboard placed over the top. One and a half gallons of *H.
ulmarius* spawn grown out on alder sawdust was added to each barrel
and evenly layered every four inches. Each barrel had approximately 20
gallons of soil added into it. No further bioaugmentation was used. At
three-month intervals these barrels will be tested for levels of toxicity,
and test results will be posted at FungiForThePeople.org.

Fig. 6.9: *In with the
white elm* (Hypsizygus
ulmarius)!
Credit: Ja Schindler

## Project: Mycofiltration of Wastewater Location: Eugene, Oregon, USA at Fungi For the People MotherPatch

At my home in Eugene I employ a very basic greywater system under
my kitchen sink: the water goes from the normal sink drain into food

grade buckets that I change out under the cabinet. If I owned the home it would drain outside, but one of the great things about this approach is the fact that I must haul the water outside, so there is another big incentive not to use a wasteful amount of water at all.

Once taken outside I use the water for various projects, usually related to mushroom growing or gardening. My problem is that there is usually a buildup of oils, food scraps and excess nutrients that would be harmful to most plants and many of the fungi that I am growing. This is a small household problem, but it parallels the same issues encountered in all food processing industries. From potato processing to ethanol production, and from biodiesel processing to community kitchens there is a water waste produced which is too heavy in mineral and oil content to be applied directly into agricultural systems. As discussed in other parts of this chapter, fungi can offer their services as biological filters and hardy decomposers of a wide assortment of organic toxins and materials. Plants and bacteria certainly have important roles in greywater systems, but oftentimes there is a sludge buildup found in greywater treatment systems, as well as blackwater systems. If the system is not designed with this in mind it can often be difficult to remove these oily buildups. What I propose is to filter them out of the water before they end up further in the system and disrupt the balance.

The following approach is designed around the idea of suspending the heavier particles of plant materials, oils and mineral buildup in a matrix of woody materials that have a healthy network of mushroom mycelium running through them. This effectively acts as a water filter, and if used in a gravity-fed design where the water is directed in a variety of directions through the filter it will be aerated and more likely to let go of its suspended material. The outflow water will in turn be lacking the usual buildup and be more oxygenated, allowing for better absorption further down the line.

This concept has many implications, and the design can be adapted to suit your situation. Another option with heavily polluted waters is to cycle the water through the system multiple times, until it exits in the desired condition. This is one of many applications currently in development of working with fungi to clarify water. There are cases where bacterially contaminated water has been safely filtered in this type of design. I would recommend being proficient in testing the water if you are intending to drink it directly upon outflow.

Before we jump into this wonderfully basic approach, I would like to put out the idea that water exiting this system will be carrying the effective enzymes of the fungi used in the design, and care should be taken to work with bacteria-friendlier fungi such as the king stropharia (*Stropharia rugosoannulata*) and the white elm (*Hypsizygus ulmarius*) as they will have a less destructive impact on the beneficial bacteria found later on down the line of most greywater systems, wetlands and watersheds.

This sample supply list can be adapted to the size of your project.

+ food grade buckets and/or water barrel cut in half
+ enough hardwood wood chips and sawdust to fill the buckets/barrels, presoaked (important)
+ sheet of cardboard
+ mushroom spawn, enough to equal at least one tenth of the woody material volume (spent spawn is fine)
+ scissors
+ marker
+ support to hold the project up (I usually just drain into more buckets)

**1ˢᵗ Step:** Remove one of the drains from the barrel, or drill some ½-inch holes in the bottom of the buckets.

Figs. 6.10 and 6.11 : *Supplies (left) and drainage (right).* CREDIT: JA SCHINDLER

**2ⁿᵈ Step:** Outline the shape of the containers on the cardboard and cut out three to four sheets for each one.

**3ʳᵈ Step:** Perforate the cardboard by cutting slices into the sheets with scissors. I usually fold them in a line and cut slices every three to four inches.

**4ᵗʰ Step:** Cover the holes in the bottoms of the containers with a layer of the perforated cardboard.

Fig. 6.12: *Careful now!* CREDIT: JA SCHINDLER

Fig. 6.13: *Perforation.* CREDIT: JA SCHINDLER

Figs. 6.14 and 6.15: *Drain cover.* CREDIT: JA SCHINDLER

**5ᵗʰ Step:** Pour in a three-to-four-inch layer of the wood chips; in bigger containers I prefer to make them lay on an angle away from the drain. Make sure all materials used are soaked well, as dry materials will damage mycelium.

**6ᵗʰ Step:** Cover this layer with some of the mushroom spawn (in this case we are working with white elm). I also like to mist the spawn to make sure it is wet as well.

Fig. 6.16: *Add wood chips.* CREDIT: JA SCHINDLER          Figs. 6.17 and 6.18. CREDIT: JA SCHINDLER

**7ᵗʰ Step:** Cover with a wetted sheet of the perforated cardboard.

**8ᵗʰ Step:** Repeat steps five through seven until you have made three to four layers of sandwiched materials. With the bigger containers it can really help to make each layer orient in a different, unlevel angle. What this does is force the water through more filter material, instead of just dropping down the middle or along the sides. In small buckets I often divot the middle of the layers in a little bit to discourage the water from just flowing down the inner walls of the bucket leaving unfiltered.

**9ᵗʰ Step:** Let the project mature by either leaving out in the rain for a few weeks, or during dry times remove the top layer of cardboard once a week and mist the under layer until water drains out of the bottom. In extremely dry weather this may need to be done daily.

Figs. 6.19 and 6.20.

CREDIT: JA SCHINDLER

**10ᵗʰ Step:** When healthy growths of white mycelium appear at the top it is ready to use. All that needs to be done is to pour cool-warm wastewater into the top and it will filter through to the drains at the bottom. Cycle through until the desired effect is achieved.

Figs. 6.21 and 6.22: *Mature filter (left) and active filtration (right)* CREDIT: JA SCHINDLER

Upon the first few runs of water through these filters there is usually a discharge of reddish water. It is the tannins which may be leaching out of the wood. If under heavy use you will also find oils from the wastewater on output. To address this second problem, cycle the water back through at a slower rate which the filter can handle.

After enough time and use the fungus will either run out of food and fruit mushrooms, or simply be exhausted and not able to filter effectively. It is recommended to treat this as any other living filter and take concern for its well-being by not exposing it to extreme temperatures nor pouring hot water though. If you are using these methods for petroleum-contaminated water or otherwise chemically toxic wastewater, I would not advise eating the mushrooms, as they will likely be composed of heavy metals and other potentially unhealthy compounds. It is best to simply leave them in the filter to let them produce new genetic variations of themselves more suited to the toxins you are working to degrade.

The decomposed contents of this system can be left to mature in the vessel, over time breaking down the filtered material even further, or transferred to suitable sites dependent on what has been filtered. In my kitchen applications the mature composting material is simply added to my compost pile. Either way, great care should be taken with the contents of this filter, and the water which drains from it, as mishandling can lead to further toxic contamination.

## Cited Works

M. Cato, ed. *Environmental Research Trends*. Nova Science, 2007.

D. D'Annibale, F. Rosetto, V. Leonardi, F. Federici and M. Petruccioli. "Role of Autochthonous Filamentous Fungi in Bioremediation of a Soil Historically Contaminated with Aromatic Hydrocarbons." *Applied and Environmental Microbiology*, Vol. 72 (January 2006), pp. 28–36.

T. Hazen et al. "Why mycoremediations have not yet come into practice" in *The Utilization of Bioremediation to Reduce Soil Contamination*, pp. 247–266.

V. Šašek, John A. Glaser and Philippe Baveye. *The Utilization of Bioremediation to Reduce Soil Contamination: Problems and Solutions*. Nato Science Series IV, Earth and Environmental Sciences Vol. 19. Kluwer, 2003.

H. Singh. *Mycoremediation: Fungal Bioremediation*. Wiley-Interscience, 2006.

## Resources

FungiForThePeople.org (courses, workshops, mushroom cultures, tinctures, consultation and research); AmateurMycology.com (Colorado-based mushroom research collective); botit.botany.wisc.edu/toms_fungi/ (Tom Volk's fungi webpage); GaryLincoff.com (internet home of mycologist Gary Lincoff); Fungalgenomes.org (home of the Hyphal Tip, an updated collection of mycological research).

*For more information on **Ja Schindler** and his organization Fungi For the People please visit FungiForThePeople.org. Ja offers courses on mushroom cultivation and mycoremediation applications; as well he sells mushroom cultures and medicinal extracts. For the past decade he has been studying mycology, and has lived and worked on a progressive mushroom farm as well as experimenting with remediation* *in small-scale industrial processes, human activity waste, asphalt degradation/removal and soil contamination. His upcoming book on mushroom cultivation will detail home and community scale mushroom growing for food, medicine and mycoremediation.*

## The Amazon Mycorenewal Project (AMP) and Chevron's Toxic Legacy: An Interview with Mia Rose Maltz

The Amazon Mycorenewal Project (AMP) is a grassroots project that is experimenting with using mycoremediation to cleanup toxic oil pits left in the Ecuadorian Amazon. Over a 30-year period, the oil giant Texaco (now owned by Chevron) discharged more than 18 billion gallons of toxic oil waste into hundreds of unlined pits in the Ecuadorian Amazon, resulting in extensive environmental contamination and health impacts for local communities. Though Chevron has recently been found guilty after an 18-year-long court battle, the corporation still refuses to pay to clean up its mess. I interviewed Mia Rose Maltz of the Amazon Mycorenewal Project about her experiences on the ground. Below are some highlights from our interview.

*Q: What do your mycoremediation experiments involve?*

**Mia Rose Maltz:** We have found some mushrooms that seem able to bioremediate the toxins; we are doing some *bioassays* that show that these fungi reduce the toxicity of contaminated soil. We've also been doing these ecological surveys that are a little bit different from mycoremediation; we go out and observe what fungi are capable of living and thriving in these heavily contaminated environments.

Our experiments involve three different tiers. One tier is *in situ*, where we do the remediation right there on the site. Then we do two different types of off-site remediation. One is open-air pits, where we dig out the contamination and mix it with a bulking agent, some sort of substrate. And then we put it into a pit lined with pond liner and then see how this kind of mycoremediation installation performs in the

Fig. 6.23: *A typical oil pit: thick, black petroleum and an oily sheen cover the ground of otherwise pristine rainforest. Thousands of unlined waste pools like this one can be found in the Ecuadorian Amazon, although "remediation" by the company is completed.* CREDIT: MEGAN SZROM

Fig. 6.24: *A student with AMP holding the soil of the jungle floor which is heavily contaminated with crude, while conducting a biological survey of plant and fungal life in the 25-year-old petroleum pools.* CREDIT: MEGAN SZROM

Amazonian environment, being in the open air, exposed to heat and other localized variables. Usually, we put shade cloth or a roof over these open-air pits. Finally, we conduct off-site experiments in boxes, like milk crates, to test specific hypotheses in order to gather data with enough replication, randomization and statistical power that the data collected could eventually be published and presented to the greater community.

*Q: How are the mycoremediation experiments going?*

**Mia Rose Maltz:** Even though we've continued to do some *in situ* experiments, we've gotten a little discouraged because some of our *in situ* installations have been tampered with, either by the military or by the oil companies. We've worked long days putting in these installations, and then months later we received word that an oil company had bulldozed it or dumped a bunch of fill soil on top of it. I really feel that the more valuable *in situ* experiments will require a long-term ecological research site that doesn't get tampered with. That is one of our big goals, to be able to buy land or have a long-term lease on land so that we could set up long-term experiments.

That's pretty much where we are at right now with everything. Still even to this day where Chevron has officially lost the lawsuit, the land that is tied up in the lawsuit is still tied up because Chevron has said that they are not going to pay. We have a really difficult situation because we have people who have had their land contaminated. We have people who have basically signed papers that have given away their right to their land to the corporation that contaminated it. People would love us to come in and do remediation on their land, but we can't because it is locked up in the lawsuit and we've been asked not to tamper with the land involved with the lawsuit. It's a really tricky situation.

*Q: You mentioned bioassays earlier. What are bioassays and why does AMP use them?*

**Mia Rose Maltz:** We've been putting some energy towards doing plant bioassays, in order to test the efficacy of the mycelial treatments, because we've spent tens of thousands of dollars on laboratory analysis

Fig. 6.25: *Mushrooms growing from oily* ex situ *boxes.* Credit: Mia Rose Maltz

and often the results have been inconclusive. The plant bioassay we've been developing is cost effective, and the results are easily interpretable. You plant 100 seeds in contaminated soil and compare the viability of the germinated seedlings with seeds planted in a soil that just has an added substrate and also soil that has an added myceliated substrate. You compare germination frequencies, the height and girth of the plants and the biomass of the plant. Then you get a sense that the plants that grow better (taller, thicker, germinate more) are indicators that the mycelial treatment has prepared the soil to be more fertile and to be capable of supporting plant establishment in previously inhospitable soil.

*Q: Why are you doing ecological surveys, and how does that fit into grassroots mycoremediation?*

**Mia Rose Maltz:** One of the things that I'm really interested in right now is looking at a community approach — the microbial community. Looking at what assemblage — what group of different types of fungi, bacteria and other microbes in combination perhaps with plants — are most effective in remediating the soil. One of the ecological surveys we're doing is a petroleum pollution gradient. We start off right at the contamination or the plume, and then move away from it, isolating and comparing the microbial community across multiple sites. We are evaluating their functional diversity, functional redundancy, or their *response diversity*, because the microbes there could potentially respond to the disturbance of the oil contamination in a diversity of ways. Some may directly perform some type of function that prepares the land for rehabilitation, while others may facilitate others to perform similar or diverse functions. Once we've looked at those characteristics, or microbial traits, then we might be able to determine if certain groups we often find occurring together might complement each other and strengthen the resilience of this ecosystem faced with a concentration gradient of surface petroleum pollution.

*Q: Are you experimenting with any other grassroots bioremediation tools besides mycoremediation for cleaning up the oil onsite?*

**Mia Rose Maltz:** This summer we did a big experiment looking at the combination of oyster mushrooms and actively aerated compost tea. We have been training our Ecuadorian partners in how to make the aerated compost tea and apply it once a week and once a month. In addition to our untreated contaminated controls, we've looked at each of these fungal and microbial treatments on their own, and the combination of the treatments. One area we focused on was the order of introduction, so basically what works better. Is it better to add the mycelium and the tea at the same time? Or is it better if we add the mushrooms and then once they are established and you have a very myceliated contaminated substrate, then we add regular applications of compost tea? Or is it better to regularly apply compost tea for several weeks or months and then add mycelium after that?

*Q: Why does the Amazon Mycorenewal Project focus so much on scientific experimentation?*

**Mia Rose Matz:** We are just trying to figure out some of the finer points with the work. What works best when you are looking at a spill in a waterway? What works best when you are looking at a spill at high elevation versus low elevation? Or really dry conditions? Or at different times of year? That's why we've been focusing on the science. We don't want to come in and say we have a panacea that will cure all your problems. That is probably the most unconscious approach we could take. We are being thorough and trying to evaluate stuff like, is bamboo a better substrate than sawdust? Are hearts of palm wood better than corn straw? Basically what is the best way to do it all? We're working with local people to figure this out, to make the greatest impact.

*Q: What tips would you have for the grassroots mycoremediator?*

**Mia Rose Maltz:** Be able to tap into who is there at the site of your contamination: which indigenous microbes and fungi are capable of surviving at the contaminated site. I would do a soil sample, add sterile water and then separate a little filtrate. Basically put some water in there and try to take an eyedropper's worth of that water and try to grow out some of those microbes and see who is there. You could do this with petri dishes. I would really encourage people to try to see who is already growing in the area where there is a lot of contamination, and also I would try some of your usual suspects. There are always a lot of cultures going around of oyster mushrooms, turkey tail, reishi; some of these fungi have been shown to be good at remediating contamination. This is also petri-dish work, however then you can get the cultures to grow on more common substrates. People need to not

Fig. 6.26: *Before image of AMP bioremediation experiment. The experiment evaluated how well* Pleurotus *(oyster mushroom) would grow on petroleum-contaminated soils in open-air pits lined with pond liner.* CREDIT: MIA ROSE MALTZ

Fig. 6.27: *After image of AMP bioremediation experiment.* Pleurotus *(oyster mushrooms) grows on oily substrate in open-air pits in Ecuador.* CREDIT: MIA ROSE MALTZ

be afraid of integrating some lab culture into their little myco group, because it really isn't hard to cook up some media like agar, fill some mason jars with it and use a pressure cooker to sterilize the jars.

*Q: How is this sort of petri-dish work helpful to grassroots bioremediators when dealing with contaminated sites?*

**Mia Rose Maltz:** I would encourage folks to do that because then they can isolate some cultures and then mix some of the contaminant in. This allows you to find a strain of fungi that is best suited to working with the contaminant and then acclimating it to maximize its effectiveness. If it's a chemical contaminant, you can soak a little piece of paper (the size of a hole punch) in the contaminant and put it in the petri dish and see if you can get your oyster or turkey tail mushroom strains to grow towards the contaminants. I would do lots of replicates: at least ten dishes with oyster and the contaminant and ten dishes with turkey tail and the contaminant.

I would say maybe one out of the ten of the oyster strains will grow towards the contaminant, and even seem to develop an affinity for that contaminant; that is the culture you want to use as your mother culture and the one that you will expand out to make spawn. So you could ramp it up, let them grow on higher and higher concentrations of contaminants, and then slowly ramp it down. With this, you are basically acclimating your strain to be able to grow in this really inhospitable environment, picking the strain and the culture that appears to have an affinity to the pollutant, and then grow on a more dilute concentration of the contaminant, in order to maximize its efficacy in more realistic applications of mycoremediation.

*Q: Besides the sterile lab procedure, what are some other options for grassroots bioremediators?*

**Mia Rose Maltz:** At the other end of the spectrum is very low-tech myco-permaculture, using your on-site resources and organic materials that are waste streams of others industrial processes. You can get some decomposer (saprotrophic) fungi to start growing on wet cardboard or other biodegradable organic materials. Another good strategy is to partner up with local mushroom growers and get a bunch of spent spawn. Bulk it up with substrate and get it growing on cardboard, sawdust, coffee grounds or other litter. And then use that material en masse, like just really go large, high spawn high inoculation, lots of it on the site, keep it moist and out of the direct sunlight if possible. Then you would have a good chance that your mushroom cultures will take.[9]

*For more information about the **Amazon Mycorenewal Project** (**AMP**), check out their website at: amazonmycorenewal.org and their website of their collaborator and fiscal sponsor, the Cloud Forest Institute: cloudforest.org/projects/amazonmycorenewal/. AMP offers field courses, internships and work exchanges in Ecuador, an annual free Shroomposium in the USA, and if you feel like donating, they are doing neat work and could definitely use support.*

## Mycoreactors

Robert Rawson and Mia Rose Maltz from the Amazon Mycorenewal Project have received both a device and a method patent for a concept called the *Mycoreactor* which they developed while down in Ecuador working to remediate the oily mess left behind by Chevron/Texaco. This simple technology allows mycoremediation to happen in soils where contamination is running deeper than mycelium normally run. Remediators dig holes down into the soil and place tubes in those holes to enable oxygen flow, along with mycelium and substrate. The tubes can be made out of wood or cardboard, making them consumable by the fungi. The airflow can be passive flow, or you can use a pump.

"Say you had a vineyard that had been sprayed with fungicides for years and years and you wanted to revitalize the soil. You could use the mycoreactor, culture soil-based fungi really easily and have them add mycelia to the soil and make it healthy again," explained co-creator Robert Rawson. "Or you could take petroleum-contaminated soil and bring the effective environment horizon deeper than what is normally available to fungi. Fungi can normally only go about three feet down because they are aerobic (oxygen loving). The tubes could be 6 feet, 9 feet or even 20 feet long, and if you are bringing oxygen down into that horizon, along with the food supply and that organism, you can anticipate that the fungi will expand its borders."[8]

## On Mycoremediation: An Interview with Paul Stamets, D.Sc.

Paul Stamets, D.Sc. is the leading mycorestoration visionary and author of several guidebooks on everything from how to cultivate gourmet and medicinal mushrooms to mycoremediation. With his many mycorestoration projects, resources, experiments and seminars, Stamets is constantly pushing the edge of what is possible when it comes to the healing forces of fungi.

*Q: What are some of the things you are working on at the moment with mycofiltration or mycoremediation?*

**Paul Stamets:** We have several projects in Mason County, Washington, USA using burlap sacks for filtering greywater. We try to find choke points where there is confluence, where we can have the maximum effect by putting mycelium at these points. Then we are able to capture contaminants and ameliorate the impact downstream of those choke points. The water tends to carry more than just one contaminant, so it is not uncommon for the water to have *E. coli*, pesticides, nitrates and phosphorous (for example). This is where mycoremediation and mycofiltration offer some unique advantages. oyster mushrooms will not only break down petroleum-based contaminants; they will also capture and eat *E. coli*, a fecal coliform bacterium, so you get a two-for-one with that species.

The more sophisticated approach would be addressing the different types of contaminants species-specifically — which means we would put a serial number of species together. You can imagine one row of burlap sacks filled with oyster mushrooms, at the front, to capture petroleum products as well as *E. coli*. If there was a mercury output from an upland source, then turkey tails have been well demonstrated to bind up mercury and mercuric ions in water with the selenium that the mycelium traps. The selenium and mercury come together form a biomolecular bond or unit that is totally non-toxic. That is one simple example where you could use oyster mushroom and turkey tails serially and then you are also using and amplifying indigenous species. These two mushrooms are prime candidates as they literally occur in

Fig. 6.28:
*Mycofiltration
installation.*
CREDIT: PAUL STAMETS

every woodland in the world. They are circumpolar — from the trop-
ics to the boreal forest up north.

*Q: How can we get more mycoremediation work happening at the com-
munity level?*

**Paul Stamets:** Every community should have a gourmet mushroom
farm — to help build carbon in the soil, to provide local healthy food
and to be able to recycle very proximate sources of debris and waste.
Every gourmet mushroom farm (they should all be certified organic)
should be reinvented as an environmental healing center so that the
mycelium can be used for remediation locally. Moist mycelium weighs
a lot; so shipping tons of mycelium across country does not make
any sense for remediation. With the debris fields that are close to the
problems, you want to keep that distance as short as possible and site

the farms in close proximity. My dream is that there would thousands upon thousands of small mushroom farms spread across the world that would be tied in to healing art centers, schools, to teaching environmental sciences, to teaching basic biology and the role of fungi in nature.

*Q: What are some ways that fungi can be used to help clean up oil spills in water?*

**Paul Stamets:** I recently invented *Mycobooms*, which are floating booms of straw filled with oyster mushroom mycelium. They can be used to corral and hold in oil and in the process of digesting the straw, the mycelium produces enzymes that break down the oil. These Mycobooms are totally biodegradable, using hemp socks that are about 20 feet in length 12 inches in diameter. They can float for three to four months.

The booms begin the enzymatic breakdown of the oil, especially the more complex heavy hydrocarbon rings; these are called polycyclic aromatic hydrocarbons (PAHs). The mycelia break them down in a stepwise fashion into smaller and smaller aromatic rings that make the PAHs then available for bacteria and other organisms to do their job too. So these fungi are the gateway species. There is a big take-home message here: These primary saprophytic mushrooms begin the sequence of decomposition that allows for a bloom and burst of biodiversity to occur so that other members in the ecological community can then use their skill sets to further break down the toxic waste. So these Mycobooms could be a gateway invention, and once you get them involved, habitat restoration occurs much more quickly.

*Q: What are some methods for mycoremediating oil spills on land?*

**Paul Stamets:** A method for land oil spills resembles sheet mulches. Layers of straw and wood chips inoculated with mycelia, 4–12 inches deep. Another extremely interesting and promising thing is that after a mushroom farm produces all the mushrooms, the substrate may be more valuable than the mushrooms themselves in terms of the economic value of its inherent enzymes. You can squeeze the enzymes out from the substrate, and you end up with this yellowish fluid that is extremely active at breaking down toxic waste. Like milking a cow, you could in a sense milk a mushroom farm, collecting the extracts coming from the substrate after it stopped producing mushrooms. Within that juice is an extremely powerful number of enzymes that can be very helpful in mycoremediation.

*Q: Any final mycorevolutionary thoughts?*

**Paul Stamets:** We need a tidal change in consciousness, and fungi offer so many solutions that we can put into practice. But it is going to take a mycological revolution on an order of magnitude such that kids learn about fungi in elementary school and in middle school. So that students and the next generations grow up to be mycologically astute, understanding that we can repair the damage we inflict upon nature. If we don't, we are shooting holes in our lifeboat; we will not only be the cause of major extinction, but we will become its victim.[10]

*For more information about **Paul Stamets, D.Sc.** and his work, please visit his website (fungi.com). His organization, Fungi Perfecti, offers mushroom cultivation and remediation seminars, resources, and you can order mushroom cultivating kits, spawn and books online. Stamet's most recent book,* Mycelium Running: How Mushrooms Can Help Save the World,[4] *is a foundational resource to read for anyone wanting to get involved in mycorestoration. His other two books,* Growing Gourmet and Medicinal Mushrooms *and* The Mushroom Cultivator, *are also great guides to help you cultivate and understand the many different types of edible and medicinal mushrooms.*[11]

For more information on Paul Stamet's take on the mycoremediation of oil spills and of the Fukushima nuclear disaster, please read:

Paul Stamets. "The Petroleum Problem." Fungi.com website, June 3, 2010. [online]. [cited November 14, 2012]. http://fungi.com/blog/items/the-petroleum-problem.html
Paul Stamets. "The Nuclear Forest Recovery Zone: Mycoremediation of the Japanese Landscape After Radioactive Fallout." [online], [cited November 14, 2012]. coalitionforpositivechange.com/stamets-fallout-mycoremediation.pdf.

## Shroomfest and Mycoremediation for Drylands — An Interview with Grassroots Mycologist Scott Koch

Based in Colorado, Scott Koch is the director of the Telluride Mushroom Festival, known by most mycology lovers as Shroomfest. Though most folks outside the region would not think to rank Colorado high on the list of fungal cornucopias, the San Juan Mountains that surround Telluride produce quite a bounty of different fungal species as a result of summer monsoon rains. According to Koch, Shroomfest covers the areas of cultivation and remediation, entheogens and medicinals, and culinary, culture and ID. This year, the festival will also offer a course in mycoremediation. With a 32-year history, the four-day event is one of the oldest mushroom festivals in the USA.

Besides organizing the Telluride Mushroom Festival, Koch plans to work with the town of Telluride on a few mycoremediation projects. One of these projects would place mycoremediation installations to reduce impacts of soil erosion and compaction at key sites where the town's many festivals take place. Another project would install myco-filtration systems (myceliated straw bales or wood chips) to intercept contaminated runoff from the town's streets. "What I want to do is create filtration systems at the places where these contaminated runoffs go before they make it to the river, which is at the bottom of the valley. All this water flows straight down from town every time it rains or snows," Koch stated. In addition to all that, Koch's dreams of creating a recycling program in town using mycelium to break down a lot of the carbon waste the town generates and to use that to build soil (which is a challenge in his environment).

When it comes to his own backyard, Koch uses mushrooms for restoration and regeneration. "Where we live is called Placerville — it's where all the mining took place in the valley. What we wanted to do was create soils and deal with the compaction that has happened over time. We are trying to create aeration in the ground by creating a matrix of myceliated wood chips throughout the property and allowing some of the pioneer species to come in and break up the compacted areas," explained Koch. "We are also on a south-facing hillside. So even in the winter when it's really cold outside, on a sunny afternoon the snow melts off. Without the myceliated wood chips there, we would have extreme erosion. Our water comes in the form of really big intense thunderstorms, and it carries away the soils we do have. We are trying to build a garden, reduce soil erosion and increase the nutrient value by building soils with rapidly decomposing saprophytes. We are using a strain of oyster mushroom and king stropharia mushroom."

Koch is also exploring the idea of heating his greenhouse using decomposing fungi: "I'm trying to do a project right now where instead of using thermal mass (like water) in a greenhouse in the winter to keep the temperature from dropping too much, I want to have the greenhouse heated by the decomposition of the fungi. I've dumped several yards of wood chips in there and inoculated them with oyster mushrooms and king stropharia in hopes that the walls will be heated by the king stropharia mycelium, and the floor will be heated by the oyster mycelium."

Koch faces some significant challenges when implementing mycoremediation in his dry and high altitude environment. "This is a bit of a touchy subject, but frankly I don't think growing mushrooms in this environment is the right thing to do. Unless you are growing a huge amount or willing to work with the natural fruiting cycle, growing mushrooms in this environment is not sustainable because it requires too many inputs. However, growing mycelium will work, and it doesn't really take too much of a different approach," he explained. "Out here, the environmental factors that you have to take into consideration are pretty extreme. There is extreme cold and extreme heat, and usually in the same day. And there is an extreme dry here; because of the altitude we have less ability to hold moisture in the air. So because of that, you need to consider where your shade is, how much sunlight you are going to get, how much water do you need and how you will maintain moisture so that things don't get fried." To help give his mycelium shade and to moderate the environmental factors, Koch also adds plants into grow holes dug in the myceliated wood chips.

When asked about which fungi species should be avoided in remediation installations, Koch had the following advice: "For beginner remediators, only use local strains of mushrooms that you know and that have been proven to be effective for what you want to do. And avoid parasitic mushrooms. Avoid anything that is a honey mushroom. They are so aggressive at killing trees and are such a parasite. It could be really detrimental to a forest if you put it into the wrong landscape. We also have this one fungi called *Pholiota* — another that is really aggressive — and I'm a bit hesitant to work with it. Anything that you are unsure about, I wouldn't consider for use."

Like many of the grassroots mycoremediators I've met, Scott Koch is dedicated to sharing his fungal knowledge and finding ways to make cultivation and remediation skills accessible at the community level. "I think that mycelium really have the potential to help with a lot of the world's issues — whether it's as preventive medicines, fighting cancer and creating new medicines, remediating the environmental issues we have or just making people happy because they like mushrooms," Koch affirmed. "I think that we have a lot to learn from these wonderful little creatures, and I think that sharing this knowledge is imperative."[12]

*For more information about the Telluride Mushroom Festival, check out its website at shroomfest.com.*

CHAPTER 7

# The Art of Healing Water

by Heather Hendrie

> *Water is fluid, soft and yielding. But water will wear away*
> *rock, which is rigid and cannot yield. As a rule, whatever is*
> *fluid, soft and yielding will overcome whatever is rigid and*
> *hard. This is another paradox: what is soft is strong.*
>
> — Lao-Tzu

In this manualfesto that speaks of Earth, we must also speak of water, without which none of the rest would exist. Water can heal us, soothe us and, from the amniotic oceans of our mothers' wombs, bring us to life. Whole chapters could be written on any one of the sentences and paragraphs that follow. Allow this chapter to pique your curiosity, guide you to courageous geniuses in the field of restorative, radical, healing water work and leave you at a few trailheads for further thought and action. Here are a couple of how-tos and next steps!

The predominant approach to water over the years has been to bend it to our purposes by attempting to control its cycles, to straighten and dam its courses or to bury it underground. We now rush water as quickly as possible through most of our communities. In so doing, we have actually engineered water scarcity into our landscapes. We have been designing landscapes that don't hold water. What becomes of a landscape that cannot hold water? It becomes a desert, the fastest growing landscape in the world.

Craig Sponholtz is an inspiring agroecologist and watershed restoration specialist. He leads cutting-edge workshops on passive water harvesting and erosion control that build skills amongst his students, who are taking on restoration projects in their home watersheds. When asked what one action we must take today to heal our waters, Craig explained: "Before we worry about regenerative work, we must strive to protect intact systems and maintain their integrity."[2]

This belief is evidenced in Craig's Guiding Principles for Watershed Restoration which include:

1. Protect and expand moisture storing areas of the landscape.
2. Stabilize active erosion and prevent further degradation.
3. Restore dispersed flow and increase infiltration at every opportunity.
4. Cultivate regenerative plant communities to build soil.
5. Create site-specific solutions using natural forms and processes.[3]

Sponholtz emphasized, as in point five above, that there is no way to be prescriptive in this work. Solutions are all site-specific, and the art of regenerative restoration lies in a deep relationship between the healer, the land and the water. There is also an urgency, as Craig reminds us that we are fighting to stave off desertification, "We're all

## The Principles of Water

These unchanging principles are relevant in any context and are thus very useful to us as we begin to work with water in various landscapes.

1. A single body of water always finds its level.
2. When pacified, water creates the conditions necessary to life.
3. Water always travels perpendicularly to slope.[5]

Understanding these simple principles allows you to work in harmony with water in your landscape rather than trying to fight the will of this powerful element. A common permaculture mantra for working with water in our landscapes is to "slow it down, store it and let it soak in!" This simple phrase guides us to recreate conditions seen in grasslands and mature forests that allow for rainwater to infiltrate rather than runoff. Keep this mantra in mind as we dive in deeper to examine filtration, remediation, liberation, resistance and water restoration work.

working for changes that we may not see in our lifetimes. But if we don't act now, more and more will be lost as we procrastinate."[4] Visit Craig's website (DrylandSolutions.com) to learn more.

## Rehydrating

Bringing water back into its natural cycles and flow patterns and refreshing our aquifers are key actions required of us to stave off desertification of our landscapes.

### Rainwater Harvesting

Rainwater harvesting is one of the most accessible gateways into rehydrating the land, noticing and honoring the water right at our doorstep. Beyond its value for water conservation, the conscientious harvest of rainwater has the potential to repair and rehydrate our immediate landscapes and reframe how we think and act towards water in our world. Brad Lancaster is a guru in the field of rainwater harvesting. He suggests that if there's one thing we need to do to restore balance to our waters it is to "re-spongify our built environment. Shift away from over-paving and sealing the surfaces of our soil. Instead, create sponges to absorb and use the water as it flows."[6]

### Direct Your Downspout

When you walk down the street, it is not unusual to see numerous downspouts pointing to driveways, sidewalks or other impermeable surfaces. Accelerating rainwater into storm drains in this way carries pollutants like oils, soap and debris from driveways and roads either directly into our rivers or — if we are lucky — into municipal engineered stormwater wetlands for treatment. To slow rainwater, direct downspouts from your home onto permeable surfaces like yards, gardens or towards a thirsty tree. This serves the multiple functions of allowing water to infiltrate your soil where it can be cleaned, nurture your growing plants and keep water from picking up pollutants and rushing them into the closest creek or river.

### Set up Some Rain Barrels

Another simple thing to do is to set up rain barrels or tanks. They are easy to install and allow you to collect soft, non-chlorinated water from your roof. A roof can capture large amounts of this plant-friendly liquid, which has the lowest salt content of any natural freshwater source. As a rough guide, assume that you can collect about one quart of water per ten square feet of roof for every one third inch of rain. Factor approximately 15 percent into your calculations for waste. Perhaps the best thing about a rain barrel is that it is a way of getting back in touch with water cycles and erecting a visible reminder of your reverence for water.

### Build Earthworks and Rainwater Oases

Beyond redirecting downspouts and installing rain barrels, we can rehydrate the landscape in numerous ways. We can and should use rain barrels, cisterns or ponds, but perhaps the best and cheapest place to store water is in the soil. Good soil, rich with organic matter, is naturally brilliant at storing water. Regardless of the type of soil on your site, the *Journal of Soil and Water Conservation* states that increasing organic matter by two percent doubles the water storage capacity of the soil.[7] So building your soil fertility by adding composted organic matter is also a rainwater harvesting act!

Brad Lancaster suggests that to start re-spongifying, we should begin with simple earthworks and rain gardens. Earthworks designed to harvest rainwater often look like a series of mulched and vegetated bowl-like shapes and depressions. Water flows through and can infiltrate the soil to be used by vegetation that acts as a living filter and pump. Beyond the potential uses it may have for us, this water can now seep deep underground to refresh our aquifers. Lancaster suggests that we focus our rainwater capture techniques within 30 feet of an impermeable hardscape (from patios to parking lots). Creating a visible oasis of lush life in a visible space highlights the flow of water and celebrates our relationship to it. These oases also provide a source of shade, protection and beauty. In dry times, where permitted, greywater from your home can even be redirected in this way through your yard

Use caution not to design within ten feet of a building so you don't saturate the soil under the foundation.

and rain gardens. Turn to the Greywater Harvesting page on Brad's helpful website for support in this area.[8]

### Hail to the Swale!

One rainwater harvesting technique that deserves special mention is the swale. Just as chickens support so many processes in the animal world and comfrey performs many services in permaculture, swales serve an incredible variety of integral, helpful functions in the world of earthworks, rainwater harvesting and planetary rehydration.

A *swale* is simply a level trench used to slow surface runoff to allow for infiltration. At their best, swales can take water running off a landscape and allow it to slowly sink deeply in, recharge underground springs and the soil and bring life back to parched and eroded landscapes. The berm beneath a swale provides a fertile planting bed, the swale itself provides a pathway for easy access to land, and best of all swales cost you nothing and can be dug using a shovel!

## Filtration and Remediation

Left alone, water flows freely through a landscape, carrying nourishment, flooding the earth with rich deposits of new soil, carving deep chasms and cycling from the earth to the clouds and back again. Wetlands, soil, rocks, plants and bacteria operate as the cleansing organs of the planet, purifying water as it passes through the land. It is time for us to seek to deeply understand these natural processes and remember through nature's wisdom how to heal contaminated waters, regenerate natural cycles and restore balance to systems that have been upset. We'll start with marshes, swamps, bogs and other wetlands, key players in the filtration game. They perform magic as they remove excess nutrients carried in water, slow its flow, permit pollutants to settle out and absorb particulate matter through plant roots.

The good news is that folks are beginning to wake up to the value of the wetlands that we've so readily filled, paved or drained in past. Our relationship to these important places is beginning to shift, and we are increasingly learning to mimic natural filtration processes. Small-scale examples of regeneration are proliferating and inspiring much larger examples. One such beacon is the 385-acre Shepard Wetland in Calgary, Alberta, Canada. The project began in 2006 and is now the largest constructed stormwater treatment wetland in Canada. The Shepard Wetland can capture and treat more than six million cubic meters of Calgary stormwater.[9]

It is relatively easy and affordable (particularly if you are dealing with soil that is high in clay) to construct a mini-wetland in your own backyard. Mini-wetland gardens can store, filter and clean runoff coming to your site from external sources, as well as from your roof and yard. Mini-wetlands also provide habitat for important creatures such as toads, dragonflies, butterflies and birds.

### *Get to Know Aquatic Plants — Superhero Water Healers*

Wetlands are largely able to purify water because aquatic plants, their resident superheroes, act as natural filters. Just as certain houseplants clean the air, various aquatic plants can phytoremediate heavy metals and other pollutants from water. Studies have found some wetland plants that contain more than 100,000 times the concentration of heavy metals that can be found in surrounding waters.[10] While all plants are helpful to a certain degree, some are better adapted to pollutant removal than others. Aquatic plants in healthy wetland systems also provide oxygen for helpful bacteria which can then perform phenomenal remediation work, transforming pollutants and toxins into less harmful or less available substances.

What follows is a list of selected species that perform phytoremediation functions in water. The list is not comprehensive, nor are all of these aquatic plants appropriate for all sites. Get familiar with our

### Caution!

Although many of the plants below may also be edible, do not eat plants if you are unsure of the quality of the water in which they are located or are using them for phytoremediation purposes. Through the process of phytoremediation, plants can store toxins in their roots, stems or leaves or release them as gasses into the air as they filter and cleanse water and land. This crucial function plants are performing can at the same time make them unsafe to use or eat.

aquatic allies below as building blocks for your own forays into healing water.

## Aquatic Phytoremediating Plants[11]

### Bulrush (*Scirpus* spp)

Bulrushes grow like a grass and up to ten feet in height. They vary in shades of green, with flowers just below the stem tip. Bulrush plants can remove oil, bacteria and other organics from water.[12]

### Cattail (*Typha* spp)

Cattails typically grow between five and ten feet high with a dark brown top that resembles a cat's tail (go figure!). Leaves are flat and twist on the plant. Cattails remove metals such as zinc, cadmium, lead and nitrates from the water supply.[13]

### Duckweed (*Lemna* spp)

Duckweeds, water lentils or bayroot are perennials with pale green leaves that typically exist as a mass of small plants (each less than ¼-inch across). They float or are partially submerged in wetlands or slow-moving, freshwater. Commonly viewed as a nuisance, duckweed actually performs many beneficial functions, and researchers are now exploring its immense potential both for nutrient recovery and use as food for people, animals and fish. Duckweeds are able to remove bacteria, nitrogen, phosphates and other nutrients from water, and researchers worldwide are exploring using duckweed as a clean energy source. Duckweeds are also great candidates for biofuel because they grow quickly, produce five to six times as much starch as corn and are a carbon neutral energy source as they remove carbon dioxide from the atmosphere while they grow.[14]

### Soft Rush (*Juncus effusus*)

Soft rush appears grass-like and grows three and a half feet high with green stems and upper stem half-flowers in a single cluster. In addition to bacteria and oil, rushes also remove heavy metals such as zinc, copper and cobalt from water.[15]

### Common Water Hyacinth (*Eichhornia crassipes*)

Water hyancinth is an introduced species and major invasive in some areas (in Florida, as a noxious weed). That said, studies highlight this plant as a promising candidate for phytoremediation of wastewater. Water hyacinth is a perennial plant that typically grows to a few inches in height and floats freely on its own raft of tiny air-filled sacs. Leaves are green and rounded and flower spikes produce 5–20 blue flowers that open following sunrise and last for a single day. Water hyacinth has been shown to improve the quality of wastewater polluted with copper, lead, zinc and cadmium. Decreases have also been noted to pH, total dissolved solids, conductivity, hardness, BOD (biochemical oxygen demand), COD (chemical oxygen demand), nitrate nitrogen and ammonium nitrogen, and increases have been observed in dissolved oxygen levels.[16]

### Water Mint (*Mentha aquatica*)

Water mint grows up to six inches high, has a mint-like look and aroma and bears light purple flowers. It should be planted in a container then put into the water no farther down than three inches below the waterline. Water mint can remove bacteria such as *E. coli* and salmonella from water.[17]

Now that you've met some of the champions of the aquatic phytoremediation world, start to enlist their help in healing our waters. Have fun and get to know these species as you collaborate with them in different configurations to create effective wetlands, floating islands and more.

## Phytoremediating Groundwater

How can we help nature clean contaminated water that runs a little deeper underground? Or what kind of barrier can we use to intercept, bind or transform contamination from reaching a water body like a river or lake? Trees! By using their deep roots and their high rate of uptake and transpiration, trees can help decontaminate and filter water. Trees should be planted above contaminated subsurface water (though they can only remediate to the depth of their roots) or in the path of a plume of contamination. Some trees especially well-suited to this are willow, poplar, alder and redwood.

## Rain Gardens

Rain gardens can be artfully designed, ideally with native plants, for specific climates, soils and neighborhoods. These beautiful planted depressions hold and absorb rainwater runoff from impermeable urban areas (roofs, driveways, parking lots and roads), creating spaces for water to be absorbed and cleansed. Plants act as natural pumps and filters to take up excess water. Their root systems increase infiltration of water into the earth and through transpiration they return water to the atmosphere. Rain garden plants can also increase soil permeability, redistribute moisture and support the microbial populations that biologically degrade and process pollutants. Rain gardens have been shown to reduce pollution reaching creeks and streams by up to 30 percent.[18] Rain gardens not only restore aspects of the natural water cycle and mitigate negative impacts on water of urban development, but can be powerful community-building sites and a point of pride for neighbors to share. Many exciting rain garden projects are underway, being led by individuals, communities and municipalities alike. In Canada, Vancouver, BC, is experimenting with some gorgeous examples, as is Calgary, Alberta. On the west coast of North America in cities like Portland, Oregon, and Victoria, BC, you can find rain gardens cut into curbs both downtown and along quiet residential streets, collecting stormwater, filtering it and supporting water in seeping deep into the earth.

Dive in deeper with a copy of Cleo Woelfle-Erskine's great book *Creating Rain Gardens* and his helpful website.[19] For another example, explore Melbourne Australia's Healthy Waterways Raingardens program. Their goal is to build 10,000 rain gardens across Melbourne by 2013.[20] Tap into their step-by-step guides and get started!

## Mycofiltration

Paul Stamets and other wizards of the magical mycelium underground guide us to harness the power of mycelium in water repair through *mycofiltration*. As Stamets has defined it, "Mycofiltration is the use of mycelium as a membrane for filtering out microorganisms, pollutants, and silt. ...These fine filaments function as a cellular net that catches particles and, in some cases, digests them."[21] Employing mycofiltration in our landscapes helps to remediate water and to channel moisture into our soils.

Refer back to Chapter 6 for more information on how to grow the mycelium you will need as water cleansing tools, or check out Paul Stamets' book *Mycelium Running* for ways to build mycoswales and mycofilters.

## Daylighting Streams

Many of us have forgotten the streams, creeks and springs that used to run through our neighborhoods, or the marshes upon which our homes, farms and shopping malls have been built. How many of us consider, when we walk down streets in the Pacific Northwest, that these roads once were creeks in which salmon used to spawn? Daylighting a stream is another creative action which helps us pay attention to the waters around us. To *daylight a stream* means to redirect it from a hidden state into an above ground, visible channel. Typically, the goal of such an action is to make the water more visible to the public and to restore a stream of water to its more natural state. Beautiful examples of this exist worldwide. One of the four largest streams in urban Seattle, Washington, USA, Pipers Creek, is now home to salmon, after 50 long years of absence. This was achieved through years of commitment and hard work by volunteers and neighbors who contributed over 4,000 hours of time to this project in 2003 alone.[23] It's your turn now — daylighting a stream may be beyond the scope of the grassroots bioremediator, but by mobilizing your community and engaging your municipal government, you can set wheels in motion to liberate waters buried underground. Go for a walk and find where your local creeks and springs may be hidden. Come up with your own creative way of drawing attention back to the water. Retrace the course of the water and the forgotten histories that run with it, and on the way, dream up some ways you can restore it to its original glory!

### Floating Islands

Another creative approach to healing water is through mobile aquatic clinics referred to as *floating islands*. The concept is that rafts of aquatic plants can mimic some of the services provided in nature by wetlands and operate as beautiful, floating treatment centers, filtering and cleaning water, as well as helping to revive dead zones in estuaries and lakes.

How do floating islands work their magic? These living structures engage both the processes of phytoremediation and microbial remediation. The roots of aquatic plants hold the islands together while also creating a rich habitat for algae, bacteria, zooplankton and other creatures to live. Microorganisms play a large part in dealing with nutrient pollution and other contaminants. Using the habitat created by the floating island, microorganisms create a biofilm that covers the plant roots and floating island matrix. Nutrients and contaminants in the water are either consumed and degraded by this biofilm or they are phytoremediated via the plant roots. When it comes to nutrient pollution (some common sources being agricultural and livestock run-off, municipal waste water and stormwater), floating islands have shown great promise in dealing with contaminants such as excess nitrate, phosphate, and ammonium, as well as carbon and suspended solids.

There are companies, like Floating Islands International®, that sell more sophisticated and specialized types of floating islands technology for a variety of different aquatic applications. But for the grassroots remediator on a shoestring budget with limited resources, you may be more inclined to build your own version of a basic floating island.

To build a floating island, all you need is some knowledge, a roll of plastic construction fencing, reused plastic bottles, zip ties, an anchor of some sort and some aquatic plant champions listed above. Ideal plants to use for your floating island are low-maintenance perennials that are native to your area. Avoid using an invasive species at all costs. Some other good choices, depending on your site, include duckweed (*Lemna* spp) iris (*Iris* spp), pickerel weed (*Pontederia* spp), arrowhead (*Sagittaria* spp) and watercress (*Nasturtium officinale*).

The following how-to steps below have been inspired by *Toolbox for Sustainable City Living — A Do-It-Ourselves Guide* by Scott Kellogg and Stacy Pettigrew.[22]

Getting started:

1. Take plastic fencing and roll it into a tube about a foot in diameter. Aim for a circumference of 15 to 30 feet and diameter of 5 to 10 feet. Close your tube off at one end with zip ties.
2. Fill up the tube with reused and capped plastic bottles. They need to be capped tightly so that they don't fill up with water and sink

your island. Stuff the empty bottles into the open end of the tube, and make sure that they are snug but that the tube can still bend.

3. Close off the other end of the tube with a zip tie and then make a circle or ring out of your tube by zip tying the ends together.

4. Complete the foundation of your island structure by pulling plastic fencing across the base of your ring and securing it with zip ties.

5. Use a good mix of aquatic plants to work their different phytoremediation magic. Take your aquatic plants and place them upright on the fencing in the center of the island, touching the inside edge of the bottle-stuffed tube. Cover the island surface with plants, but avoid overcrowding. Gently work the roots of the plants though the holes in the fencing so that they will hang down into the water. With large roots, zip-tie them to the plastic fencing. You want your plants to stay upright, so you may want to secure the stems or leaves of the plants to the sides of the tube if necessary.

6. To increase the amount of oxygen in the water, further supporting remediation, consider attaching a solar-powered air pump to the island with plastic hose and air stones dangling deep below in the water.

7. Pick a location to deploy your floating island, preferably in full or partial sun so the plants can flourish.

8. Launch your island by boat or by wading in from shore. Be sure to secure the island using an anchor system so that it doesn't float away.

9. Come back and check on your island periodically to make sure it is alive and well. If it is not, remove it from the water so that it does not become floating or submerged garbage that causes the aquatic ecosystem more harm than good.

As your island matures, a living system will develop and hopefully flourish, both purifying the water and providing habitat and food to the aquatic environment you are seeking to heal. Habitat will be created for insects, birds and fish. Depending on the size of the water body, you may have to construct many floating islands and consider different sizes.

Some concerns have been raised about the plastic bottles photodegrading and releasing contaminants into the water. If this is a concern for you, you could consider switching out your floating islands every several years to minimize that.

## Tensile Strength

In the polarity between ripples and stillness, flooding and drought, destruction and regeneration, issue and solution, lies a lesson for us in the diversity and resilience required to live on this Earth. Water's nurturing force, when calm, to foster and support new growth and the emergence of life is unequaled. Just as this is true, so too is her power in fast, swollen states, to clear the land, sweep away foundations, erode, drown and destroy existing structures. There is a beautiful power in both of these extreme personalities of our waters, and both are needed.

The same is true of our activism. At times we may rage like swollen rivers, to sweep through the streets in masses and uproot structures that no longer serve us. At other times we best serve the world by digging our hands into the soil and pacifying the water, allowing it to infiltrate and nurture new emerging structures. Both of these seemingly opposed approaches and everything in between are needed at this time on our planet. Whether we are bureaucrats, teachers, elders, artists or environmental organizers, we are all being called upon to come alive right now. And we live at a time when, in defense of water, we may be called upon to play several roles. There are many who are working tirelessly to ensure that certain sacred rivers and streams never get dammed or buried underground. Like the 300 Indigenous people, small farmers, fisher folk and local residents who occupied the Belo Monte Dam project in Brazil, removing a strip of earth to restore the Xingu's natural flow and "freeing the river."[24] There are those who are striving creatively to remind us that water continues to run beneath our feet. All across the world, people are rising up in defense of water and to fight for its liberation.

Every choice we make affects our watersheds. It is time for us to use our minds to steer wise action, our bodies to regenerate, repair and restore waterways and our voices to speak hard truths about the way forward together as we value, protect, fight for and lovingly heal our wild waters. Remember now that we are beings of the Earth, lovers, healers, stewards and a key, interconnected strand in a web of life that is held together by droplets of shining, sacred water.

**Heather Hendrie** *is a feisty freelance writer and champion for clean, running water. Heather believes that her body is approximately 53 percent Bow River, 21 percent Georgian Bay and 2 percent Victor Davis Pool. From swimming to canoe tripping, raft guiding and environmental education, Heather's personal and professional life has revolved around water. Most recently, Heather worked to promote the value of water through The City of Calgary's Water Services department.*

CHAPTER 8

# Oil Spills I: The Anatomy of an Environmental Disaster

THIS IS A HEAVY PART OF *EARTH REPAIR*. We dive into the crude realm of oil spills and the whole host of ecological and human suffering that they cause. For me, even though the ways to deal with these disasters are difficult, dangerous and can be far beyond the scope of a community response, I felt them important to include. Oil spills are one of the most undeniably catastrophic and visible costs of global society's addiction to oil.

The stakes are getting higher too. As easy-to-access oil becomes more scarce, companies and governments dismiss mounting risks and impacts associated with the pursuit of the more unconventional and dangerous sources of oil (e.g. tar sands, offshore and arctic drilling). Unconventional sources of energy require technology and infrastructure that are often approved prior to sufficient testing in field situations. Spill emergency response plans often look nice on paper but scarcely hold up when disaster hits. The increasingly extreme and remote environments in which oil, gas and mineral deposits are found can make spill response incredibly difficult (if not sometimes impossible), threaten extremely sensitive, rare and slow-to-regenerate ecosystems and have exponential consequences that are hard to localize and contain. Furthermore, under the auspices of tight economic times and austerity measures, many governments are cutting back resources to key disaster response and environmental emergency programs, even as they and energy corporations undertake more destructive and dangerous projects.

In the end, this is a recipe for disaster, where the consequences will fall on the shoulders of frontline communities and the planet.

So in the spirit of grassroots empowerment, this chapter and the next at the very least seek to give you more information about oil spills, how they impact ecosystems, how to keep yourself and your community as safe as possible in their wake and some small things you can do to be of assistance in the aftermath of a spill.

## Oil Spills 101

All oil spills are different, and how they are handled can make all the difference in the severity of impacts down the road, as well as the toll they take on an ecosystem. As a grassroots bioremediator, you should understand how oil spills behave. There are also many different types of spills to consider, and they all pose their own challenges for cleanup. There are spills at sea, on land, and in rivers, lakes and wetlands.

How do large oil spills happen?

- tankers leaking, colliding or running aground
- offshore blowouts
- pipelines leaking or bursting
- tanker trucks carrying fuel colliding or rolling over and spilling their contents into a nearby land or water body
- leaks from refineries and holding facilities
- corporate negligence and irresponsibility

Land spills can be easier to contain than marine spills. River spills are harder to contain due to the speed of moving water. But the water's mobility also means it is higher in oxygen and allows for bioremediation

## Chemicals to Watch Out For

In an oil spill situation, you need to be aware of what you are working with in order to understand the risks to you and the planet and properly respond. Some of the main contaminants are crude oil, polycyclic aromatic hydrocarbons (PAHs), volatile organic compounds (VOCs), dispersants, surfactants, fertilizers, detergents and heavy metals. See the Appendix 1 for more information.

to occur more naturally; oil in moving water will mechanically break down faster than in a marsh or lake, where water is more stagnant. Dealing with all of these spills involves first stopping the leak, and then doing what you can to contain and remove the oil.

## Factors Affecting Oil Spill Impacts and Cleanup
### Response and Containment Time

The quicker you respond to an oil spill, the better the chances of containing it, averting the biggest, deadliest impacts and keeping oil from hitting shorelines, beaches, moving down rivers or seeping deep into the land. The quicker and more effectively you are able to contain the oil, the quicker you can remove the oil before it starts to evaporate, weather, sink or smother plants, wildlife populations and shorelines. It is important to note that a quick response time in many cases is contingent on frequent monitoring of the site so that the spill is discovered immediately. This can be a challenge, as sometimes a company will keep a spill under wraps to avoid a public relations storm and its clean-up responsibilities. Government regulatory agencies simply do not have enough boots on the ground for effective and frequent monitoring. The more remote a project is (offshore and arctic drilling for example), the less likely it is for word of a spill to get out unless the company notifies the government and public.

With marine spills, if oil is left too long at sea and is subject to wind and waves, it emulsifies and changes, in about two weeks time, to a thick, gel-like substance called *chocolate mousse*. Yummy? Not quite. This mousse is a mixture of oil, air and water. Mousse from medium and heavy oils can cling to rocks and sand. Heavier oils, when exposed to sunlight and wave action, can form dense, sticky and toxic substances known as *tar balls* and *asphalt* that are very difficult to remove from rocks and sediments. Through wave action, tar balls can sink to the bottom or wash up on shore, continuing to leach into water and sand.

### Type of Oil Spilled

Not all oil is the same, and different types of oil respond differently in spill situations due to their different viscosity and molecular weights. Gasoline, kerosene, heating oil and diesel are considered lighter forms of oil. Bunker A, Fuel Oil No.4, lubricating oils and medium crudes are medium forms. Bunker B and C, Fuel Oil No. 6, weathered crudes and bitumen are considered heavy forms of oil.

There are existing oil spill clean-up methods for the lighter to medium oils, as these oils float on the surface of water for a while and form a sheen, allowing them to be contained and sucked up or sopped up when possible. Lighter forms of oil tend also to evaporate and degrade more quickly, so that less is deposited on shorelines and beaches. Spills of lighter oil still cause a lot of damage and can be very toxic, but they are not as difficult to clean up or as persistent as the heavier types of oil. These heavier or thicker oils, like tar sands bitumen, could either become neutrally buoyant and remain suspended in the water column or sink quickly to the bottom, making it hard if not impossible to recover by traditional containment or skimming methods.

Also, bitumen is so thick that it must be diluted with a mixture of volatile and toxic chemical solvents in order for it to pass through a pipeline. This mix is called *diluted bitumen* and it is much more toxic than conventional crude. It also has to be piped under high temperature and pressure, and when you combine that with the more abrasive and corrosive nature of the mixture, you have a recipe for a pipeline spill disaster. In the event of a spill, the solvents used to dilute bitumen quickly evaporate (releasing toxic chemicals like benzene and toluene) and pose increased public safety and health risks.

### Weather Conditions

Ideally when oil spills, especially marine spills, you want calm conditions with little wind or wave action. This allows oil to be more easily contained. It also allows other industrial techniques (e.g. using dispersants or *in situ* burning) to be used more effectively and safely by industry standards. Calm weather also makes conditions for emergency responders and clean-up crews easier. High wind and wave action makes booms ineffective and can lead to the oil becoming emulsified. Incredibly hot weather can make oil spill cleanup hard on volunteers and responders, as it makes it more uncomfortable and difficult to wear protective gear, and also leads to heat-related health issues like dehydration and heat exhaustion.

The worst case scenario would be a spill followed by big storms (dangerous marine conditions, poor visibility, heavy winds and rains)

that could scatter the oil, push the oil towards shore/into the environment, or make it difficult or unsafe for clean-up crews to work. An oil
spill followed by hurricane season or heavy rains that could lead to
flooding or landslides is definitely not something you want, neither is
a spill in forest fire season.

### Currents, Tides and Waves

For ocean or river spills, strong currents or increased wave action can
help naturally break up and break down oil. But it can also make it
hard to contain oil, push it up onto the shoreline and also cause it
to emulsify and become less manageable. It is important to know the
direction of any currents in order to determine where oil will flow and
collect.

### Geology of Shoreline and Landscape

Brackish marshes, coarse gravel beaches, coarse sand beaches, fine sand
beaches, rocky shores, salt marshes, coastal structures (consolidated
seawalls, consolidated shores), wetlands, tidal flats, mud flats, vegetated riparian shoreline — each type of shoreline and landscape will
handle and hold spilled oil differently, and some are easier to clean than
others. A marine biologist by training, Paul Horsman is the Global
Campaign for Climate Action Director for Greenpeace International.
"If you have an oil spill on a rocky coast, you can put booms around
and try to skim it off and try to remove it as much as possible without
causing more damage. But if you have an oil spill where you have salt
marshes or mangroves or low-energy shores, the very act of getting
in there can cause damage as well. For some spills in particular environments, nature does much better than we can do in helping with
the recovery."[1] Similarily, for terrestrial oil spills, oil will move quicker
through sand or gravel than it will through clay.

### Temperature

This is an easy one — oil spills that happen in warmer places and in
warmer waters break down more quickly because of increased microbial activity (most microbes are more active at warmer temperatures).
In colder climates (where there is an inhospitable winter involved), microbial action is slow, and clean-up operations usually have to stop for

periods of time due to winter conditions making the work difficult or the site and oil inaccessible.

### Sensitive Biological Populations

If an oil spill occurs during salmon spawning, whale migration, calving season or when shorebirds are nesting, the damage will be heavy. As a result, certain oil spill clean-up techniques may be used to divert the oil away from sensitive sites.

## How Industry and Government Clean Up Oil Spills

Oil spill cleanup is a resource-intensive empire, with clean-up specialists, contractors, companies and governments constantly inventing and patenting an ever more complex set of spill response technologies. Oil spills are disastrous situations, forcing folks to make rough choices and sacrifices. In the end, it is important to remember that sometimes these choices are made from weighing all the different magnitudes of destruction and choosing the least awful situation, and sometimes they are based purely on ignorant, greedy or face-saving motivations.

It is important for earth repair responders and community members to understand how industry and government handle oil spill cleanup because you will either find yourself working alongside them in a volunteer capacity, or at the very least will want to separate the spin from what is really happening so that you can either try to stop them from doing a bad job or take proper precautions to protect your health and that of those you love.

### Containment and Removal
### Booms

The first line of defense and response to an oil spill often is a mechanical one, involving containment, skimming and sorbents. Using specifically designed barriers known as booms, floating spilled oil can be contained in low-wave, low-current environments. Commercial booms are available for a wide variety of applications ranging from rivers to the offshore.

Booms float on the water surface to corral spilled oil using three parts: a freeboard that rises above the water surface, contains the oil and prevents it from splashing over the top; a skirt that hangs below

Fig. 8.1: *Containment booms in Louisiana.*

the surface and prevents surface oil from being pushed under the boom and spreading; a cable or chain that connects, strengthens and stabilizes the boom. A different type of boom can be used on shorelines as well, to stop oil from moving onto the shore or beach, though success is highly dependent on tides and wave action.

Certain situations can make it hard for booms to contain oil. When winds are stronger than approximately 20 knots and waves are higher than three feet, oil cannot be contained in a boom. Similarly, a boom is not effective in strong currents, since if the current is stronger than about one knot, oil will splash above the boom or escape beneath it. Floating booms are largely ineffective on heavier crude, which tends to sink to the bottom.

## Skimmers

Once oil has been contained using booms, a variety of different skimmers suck up the oil like vacuum cleaners or scoop oil from water into containment tanks or barges. Skimmers can get clogged easily and don't always work well on large oil spills or when the water is rough. Also, a skimmer should be suited to the viscosity of the oil in question.

## Sorbents

In some situations, either to remove contained and floating oil or to help with shoreline or land cleanup, sorbents are used to sop up the contained oil. *Sorbents* are materials that act as large sponges that soak up oil by either absorption or adsorption. Oil will coat some materials by forming a liquid layer on their surface (adsorption). Sometimes companies will put absorbent materials like hay on beaches near an oil spill or spread vermiculite over the spilled oil. Absorbent materials, very much like paper towels, soak up oil from the water's surface and from rocks, soil and animal life that becomes coated.

## In Situ *Burning*

*In situ* burning involves controlled burning of oil at the location where the oil spilled from a vessel or a facility. Typically, oil is contained within a fire-resistant boom (often a ceramic fireproof fabric or stainless steel material) and ignited. The oil sheen needs to be thick enough for it to continue to burn, which means it must be anywhere from one third to three quarters of an inch thick. Many folks in industry and government believe that given the proper conditions, *in situ* burning is a fast, efficient, inexpensive and relatively simple way to remove spilled oil from the water that also decreases the need for storage and disposal of collected oil and spill clean-up wastes. There are some situations, such as marshes or ice-covered water, where other forms of manual or mechanical oil spill removal are limited, inaccessible or may cause too much physical damage from equipment and trampling. It is believed in these situations that *in situ* burning is one of the only options. Furthermore, because it removes oil quickly, burning can be used if there is a time critical element to the oil spill response, such as upcoming flood or hurricane season, or the arrival of large number of migratory birds and mammals.

However, *in situ* burning in marshes can cause substantial damage to vegetation, mortality to organisms unable to escape, create oil residues that may be hard to remove and sometimes can result in a deeper penetration of the oil into the substrate below. Keeping the burning under control is very important when working in terrestrial situations. *In situ* burning is only effective when there are calm conditions. If winds or waves are too strong, containment will be difficult. Furthermore, there is usually only a small window of opportunity when the right conditions are present for a burn, and if that opportunity is missed this method is no longer useful.

*In situ* burning also converts oil into airborne gases. Emissions from *in situ* burning include carbon dioxide and water, particulates and small quantities of nitrogen oxides, sulfur dioxide, ketones, aldehyde, polycyclic aromatic hydrocarbons (which are known human carcinogens) and other minor combustion gases.[2] These emissions can fall back down to earth or can travel long distances away from the spill site via a toxic smoke plume.

It is true that burning oil on-site does reduce the amount spilled and can keep an oil slick from smothering a coastline. But it is also true that burning spreads pollution in different ways. Whereas responsibility for and outrage from the public for an oil slick, tar balls on a beach

Fig. 8.2: *Burning oil from oil rig disaster.* CREDIT: © DANIEL BELTRA / GREENPEACE

and oiled birds can be traced back to an oil company, it's a lot harder to connect sick people miles away with inhaling mostly invisible toxic fumes from the spill. It is critical to make sure that if burning is to take place, downwind populations and communities are notified and vulnerable individuals (children, elderly, pregnant women, people with respiratory and autoimmune illnesses) are evacuated. Workers must be given respirators and proper safety gear. All in all, the decision whether or not to burn involves some serious trade-offs for local communities and the environment that need to be properly considered.

### Letting Nature Take Care of It

In some situations, industry will decide to simply let natural forces break down the oil. Basically oil will eventually degrade and dissipate as a result of sun, waves, weather and bacteria. This is true in some respects, but it takes a really long time and can be incredibly slow in colder environments. It also causes lots of impacts in the meantime, especially as submerged oil continues to leach into surrounding land and water with serious toxic consequences for wildlife and local communities.

### Dispersants

Dispersants are one of the most intensively used methods to handle large oil spills. According to the Department of Fisheries and Oceans in Canada, "Chemical oil dispersants work much like dish soap by changing the surface tension of the oil so it breaks apart into very small droplets that mix more easily with water. Tides, currents and other physical processes help to disperse the oil into the water column (below the water surface), where naturally occurring bacteria can break down much of the oil into non-toxic compounds."[3] Dispersants are applied from aircraft or boats. Less oil on the surface also decreases the extent that wildlife will be oiled, or the oil will smother the shoreline.

However, there is growing evidence and mounting concern from scientists, environmentalists, sick oil spill workers, volunteers and impacted communities that dispersants can have incredibly toxic impacts, made even more so when combined with oil and salt water. When dispersants allow oil to be broken down into smaller particles throughout the water column, a problem on the surface becomes a problem ingested and incorporated into all living beings in the affected area.

Paul Horsman had the following to say about dispersants: "As a general rule, I would not support the use of dispersants. But with all rules there are exceptions. If you had a surface oil spill that is near to or heading for a breeding colony of birds, the dispersants could actually be useful in saving a number of birds from that breeding colony, because it can disperse the oil in the water. But of course if you have fishermen, and this was certainly the case in the Gulf of Mexico, they don't want to use dispersants, because they know that when you use dispersants, it pushes the material into the environment and actually adds to increased problems with fisheries. Also, dispersants only work on freshly spilled oil, and only certain kinds of oils. They don't work on all spills, and they don't work on oil that has been in the environment for more than 24–36 hours. In the BP Gulf Horizon spill, thousands and thousands of gallons of dispersants were injected into the oil as it was coming out to keep that horrible black sheen from getting to the surface. But that was an uncontrolled experiment being carried out because that technique had never been done before. I think we will see the results of that spill for many years to come because of the huge use of dispersants. There is a huge amount of pressure on an oil corporation to reduce visual damage and to get the oil out of sight as quickly as possible. That isn't an environmental pressure, that is a public relations pressure. Spraying with dispersants is in very few circumstances the best thing to do."[4]

### High-Pressure Hoses, Vacuum Trucks and More

If a spill reaches shore, especially a rocky shore, high-pressure hoses can be used to "wash" oil off the beach. Vacuum trucks can also be used to suck it up oiled sand or oily runoff. Sand, gravel and soil can also be dug up, using shovels or excavators, though big machinery is likely to cause biological damage, erosion and in some cases can push oil farther down into the sediment. For beach cleanup, sorbents and sponges are also used to pick up oil or wipe off oily rocks.

One of the biggest issues with beaches and shorelines is submersion of the oil. If oil sinks too far down into the sand, it will continue to sit there and seep into the environment, poisoning the substrate and impacting the recovery of species that thrive there. High-pressure hoses and hot water washes can create the appearance of a clean environment, but they can also push the oil out of sight. This method can destroy sensitive habitats and the millions of tiny species that depend on them; though largely invisible to the human eye, the destruction of these beings can reverberate through the food chain for decades to come and mean all the difference between ecosystem recovery and collapse.

According to Paul Horsman, "The classic example from a US perspective is the *Exxon Valdez*, where in order to be seen to be actually doing something, a lot of people were a lot paid money to run around washing rocks, hot water pressure washing and spraying with chemicals. All of which caused more damage, and in fact pushed the oil further into the environment, instead of actually removing it. Now it is pretty well recognized that some of the techniques used for the *Exxon Valdez* spill, indeed on most spills, caused more damage to the environment in the longer term than did the original oil."[6]

### Floating Dummies, Balloons, Water Cannons, Buoys and Horns

Companies and clean-up agencies often use floating dummies and balloons to scare wildlife away from spill area. Cannons and horns can also be used to try to dissuade birds from landing in an oil slick. Sometimes this works but often doesn't; for example many ducks have perished in Syncrudes' tar sands tailings ponds in Alberta.[7] If it is a small spill and you can put netting over it somehow, that can also keep some birds from being oiled, but this doesn't work in many instances because of the size and location of the spill.

### Big Bioremediation — Using Microorganisms and/or Biological Agents

All marine, freshwater and terrestrial ecosystems contain some oil-degrading bacteria. No one species of microorganism, however, is capable of degrading all the components of oil. Petroleum is a complex substance and therefore the microbial communities that feed on it are also complex. Many different species are usually required for significant overall degradation, and the different components of oil are degraded at different rates and extents, some quicker than others. In an ocean spill, marine-dwelling microorganisms use the hydrocarbons of the oil spill as food, emitting carbon dioxide ($CO_2$) as a result. In some places where there are frequent oil releases (like the Gulf of Mexico) or where there are natural seepages, there are more local oil-eating bacteria present than in places where there are no oil releases.

Bioremediation is done for both terrestrial and marine based spill though it is easier to do on land spills. In marine spills, chemical dispersants are often used to break the oil into smaller droplets so that bacteria can more easily consume them. Oxygenating the water using a pump speeds up bioremediation greatly. When an oil spill reaches the shore, industry can use *biological agents*: fertilizers, like phosphorous and nitrogen, are spread over the oil-slicked shoreline to boost and accelerate the rate of growth of local microorganisms that will then break oil down into natural components like fatty acids and $CO_2$. This method of bioremediation is called *fertilization* or *nutrient enrichment*. When bioremediation is done in this way, one should be careful as fertilizers can cause algal bloom and can also be toxic, especially in combination with other oil spill clean-up chemicals.

*Seeding* is another way of doing bioremediation on an oil spill. Microorganisms (either naturally grown or genetically engineered in a lab) are added to the spill site, sometimes along with nutrients to kick off the biodegradation of the oil. For decades scientists have been experimenting with genetic modifications that might enhance the ability of certain microbes to chew up oil.

## Inside the World of Big Bioremediation:
## An Interview with Anita M. Burke

Anita M. Burke is an international energy and environmental remediation expert who has worked with energy giants like Exxon, Texaco, Shell US and Shell Canada. Burke got the opportunity to do bioremediation at a Shell spill in Fidalgo Bay near Anacortes, Washington, USA.

*Q: What do you do to bioremediate an oil spill?*

**Anita Burke:** First you've got to find out who is hanging out in that ecosystem that likes to eat decaying crude oil, and then figure out what food you are going to give them, fertilizer if you will, to get them to eat that stuff. What you are trying to do is build up the natural biological critters that are there and falsely explode their population in order to attack the food source, which is the decaying crude. They'll die off eventually when the food is gone, but you have to first create an ecosystem for them.

*Q: Were you able to successfully bioremediate Fidalgo Bay? Did the ecosystem recover post-spill?*

**Anita Burke:** Today, you can still find crude up at the *Exxon Valdez* oil spill. Because what we did with the hot water wash and the Corexit [dispersant] is we killed all the biology. It looked good and Exxon got to say that the beaches were clean, but we killed everything with hot water, steam and nasty chemicals. So nature couldn't even help itself after we left. But with Fidalgo Bay, it's different. Eight years after the spill, I'm sitting around the fire with my family and my cousin and her husband had just done a cross-country bike tour and they were talking about Fidalgo Bay. How stunning it is, how it is a protected estuary that's teaming with life. And I'm just sitting there grinning, because you would never have known that there had been an oil spill there. Because we amplified nature's ability to help herself.[5]

---

## Health Impacts of Oil Spills and Conventional Cleanup

Oil spills are overwhelming situations, and they can quickly overwhelm companies, coast guards, emergency responders and communities. When an oil spill happens, many people are compelled to help out. People sign up to volunteer, whether it is sifting tar balls out of beach sand, being part of a mosquito fleet of boats laying booms, helping spray the arsenal of chemicals that the company has instructed be sprayed, cleaning off rocks, de-oiling mammals and birds, cleaning oiled work clothes and more. But what many community volunteers and workers have learned quite tragically from working on spills such as the *Exxon Valdez* or the BP Gulf Horizon is that you cannot trust the company or the government to consider your health and safety or to properly inform you about the risks associated with the impact of the spill, the work conditions you are in, what chemicals you and your neighbors are being exposed to, the effectiveness of the cleanup and its impacts on the safety of your food and water sources. It is important to realize that when you respond to an oil spill, or live on its edge, you are being exposed to a whole range of toxic chemicals that have the potential to impact your life in very serious ways.

Like many oil spill workers, Anita M. Burke found this out the hard way. "I was literally dying from my exposures from the *Exxon Valdez* spill, like many have. The average life expectancy of an *Exxon Valdez* oil spill worker is 51 years of age. I'm 51 now, and I'm still alive. But there aren't many of us left. I got really sick, and I wasn't alone. We were given doctored MSDS sheets [material safety data sheets] and were told that things like Corexit [the dispersant used] was a non-toxic agent," Burke recalled.

When asked about what sort of health impacts she suffered, Burke described the following: "What we didn't understand at the time was that when you mix Corexit with crude oil and salt water, you create a whole new compound that off-gases butyl ethers. The butyl ethers are the things that give you an acute exposure response that looks like the flu or a cold. Your chest is all congested, you are throwing up, and it looks like a really bad winter flu. That's the acute exposure response. And it doesn't flush out of your body. The more you are exposed, the more it accumulates. Then it slowly but surely eats away at you. So five to ten years later, all of a sudden you can't get out of bed, your legs aren't working, you are having stroke-like symptoms, you are staggering, drooling, your brain isn't functioning right. You have brain fog; you don't remember where you are. I went to doctors to figure out what was wrong with me and they didn't know, but they told me my endocrine system was failing and that I had six months to live. I ended up starting chelation therapy and alternative therapies. I was super sick for two and a half years, and the chelation therapy saved my life. It's been hell. But I'm OK now."[8]

In the aftermath of the BP Gulf Horizon spill, many stories have surfaced of clean-up workers, volunteers and community members falling ill with serious symptoms that are really hard to treat, sometimes resulting in death. Some became ill shortly after their exposure, others not really showing symptoms until over a year later. When tested, people have been finding chemicals like toluene, benzene, xylene (all volatile organic compounds) in their blood. They are suffering

from rashes, sores, respiratory illnesses, neurological and nervous system issues and many other debilitating conditions.[9] Allegations have
also arisen that clean-up workers and volunteers were not given proper
safety gear, like respirators, when engaging in spill response.[10]

Many of the chemicals present in the spilled oil, dispersants and
other chemicals being used (solvents, detergents, fertilizers) are known
to cause headaches, nausea, vomiting, kidney damage, altered renal
function and irritation of the digestive tract. They can also cause lung
damage, burning pain in the nose and throat, coughing, pulmonary
edema, cancer, lack of muscle coordination, dizziness, confusion, irritation of the skin, eyes, nose and throat, difficulty breathing, chemical
oversensitivity, delayed reaction time and memory difficulties. Further
health problems include stomach discomfort, liver and kidney damage,
unconsciousness, tiredness/lethargy, irritation of the upper respiratory tract, hematological disorders and death. If that isn't a list peppered
with skulls and crossbones, I'm not sure what is. The main pathways
of exposure to the chemicals and their resulting impacts are through
inhalation, ingestion, skin and eye contact.[11]

In the case of the Gulf Horizon spill health impacts, BP is denying
all responsibility; just as their massive use of the dispersant Corexit
has caused the oil to be out-of-sight, out-of-mind, so, apparently, is
their attitude to the clean-up crews and personnel that they employed
to clean up their mess and the communities that still have to live in the
shadow of the spill.

# CHAPTER 9

# Oil Spills II: Tools for the Grassroots Bioremediator

WITH SOME OF THE BASICS ON OIL SPILLS COVERED, let's move on to what we, as grassroots remediators, can do in these situations. Big spills like a BP Gulf Horizon would put most of us in a pretty serious damage control mode, and the amount of company and government involvement would make it hard to deploy grassroots bioremediation techniques, as they are often discouraged during professional cleanups. But there is space for grassroots bioremediation, mostly in the aftermath once the company and their experts have left and gone home (if they ever came in the first place), doing shoreline cleanup or dealing with a small spill, preferably on land.

Though there have been some professional industrial bioremediation efforts and attempts, grassroots bioremediation techniques have yet to be done effectively on a large-scale or in a marine setting. However, they have been deployed on a smaller scale and in more contained situations. So really, *Earth Repair* speaks to those smaller instances, specifically land spills. However, if a large spill happens near you and there is no response besides a community level one using whatever is on hand, consider trying out the bioremediation tools described in this chapter, being sure to take proper safety and health precautions.

## Using Natural Sorbents

For the grassroots bioremediator trying to get oil off land, a beach or shoreline (and in only certain cases out of the water), there are more organic, biodegradable, non-toxic, easy to make and natural sorbents that can be used instead of commercially synthetic ones. You would deploy these either to sop up oil that has already landed on a beach or the ground, or you could try to lay them down as a damage control measure to absorb the oil when it arrives. When it comes to containment and oil collection, time is of the essence, so the quicker you can respond, the better the chances of reducing the extent of the damage. I also describe a few natural booms below, as some of these devices double as both containment booms and sorbents. Once saturated with oil, these natural sorbents and booms can be broken down using either mycoremediation or composting, or they can be disposed of as hazardous waste.

## Hair Mats and Hair Booms

Highly effective sponges, hair mats were invented in 1989 by Phil McCrory, a hair stylist from Alabama, USA. Human hair and pet hair attracts and soaks up oil; it's a resource that is easily available and cheap to produce, as most of it can come by donation. Hair mats and booms are easy to make and are a way for many people to contribute to an oil spill clean-up effort by organizing in their community to work with salons and pet grooming businesses to collect hair and to make and stuff booms.

One pound of hair can suck up one gallon of oil. The hair and fur can be made into *hair mats*, which are placed along the beach, shore or on an oil spill to sop up the oil. *Hair booms* are made from packing hair and fur into nylon (e.g. pantyhose or tights) to form long, sausage-like natural oil booms that are a more environmentally friendly than commercially made plastic booms. Once saturated, both mats and booms can be collected and either wrung out to capture spilled oil (if you do this, do it in a well-ventilated, open-air space with proper protection gear so you do not breathe in any fumes or get oil on your skin), or they can be moved to a holding tank of some sort where they can be mycoremediated and broken down. You could also compost them (in a special compost, not in your backyard garden one).

Some folks advise against using hair booms or hair mats in the water, as they could potentially sink as they fill with oil. Others refute this claim.

## Hairmats and Grassroots Bioremediation of Big Oil Spills: An Interview with Lisa Craig Gauthier of Matter of Trust

Lisa Craig Gauthier is the president of Matter of Trust, a San Francisco-based nonprofit that has garnered much attention for spearheading grassroots bioremediation efforts notably around the *Cosco Busan* spill in 2007 and in response to the BP Gulf Horizon spill in 2010. In both cases, Matter of Trust worked hard to organize and mobilize a volunteer effort to get hair mats and hair booms deployed in the spill areas. In both cases they were blocked, by the government in the *Cosco Busan* spill and by BP in the Gulf Horizon spill. Here is part of an online interview I conducted with Gauthier.

*Q: How do you make a hair mat?*

**Lisa Craig Gauthier:** Felting is the oldest textile production technique in the world. It's basically jostling and wetting hair; it's like making flat dreadlocks. But it takes machines to do it if it's going to be practical.

*Q: Is Matter of Trust still keen on using hair mats for oil spills?*

**Lisa Craig Gauthier:** We would love to use hair mats, but felting machines are hard to find in the US — most have been sold off to Asia and Mexico. The textile industry has almost collapsed in the US, mostly due to petroleum prices, ironically enough. So many textiles are synthetic, and the price [of oil] going from US$80 to US$140 a barrel just ruined small family textile businesses, and large companies have moved overseas. It has been so hard to get recycled hair felted since 2008 that we started making booms (hair sausages) ourselves instead.

*Q: How do you use hair booms?*

**Lisa Craig Gauthier:** We use booms mostly because they are so easy to make. You just stuff hair into a nylon stocking. That is what we used for the BP spill in the Gulf. We have lots of videos that show you how they work. Booms can dam and float so they can be for land, marsh or water.

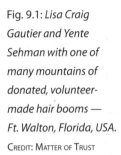

Fig. 9.1: *Lisa Craig Gautier and Yente Sehman with one of many mountains of donated, volunteer-made hair booms —* Ft. Walton, Florida, USA.
CREDIT: MATTER OF TRUST

*Q: Can you describe the mass grassroots mobilization that happened to get hair mats and hair booms to remediate the BP Gulf Horizon spill?*

**Lisa Craig Gauthier:** The EPA (Jared Blumenfeld, EPA Administrator Region 9) called it "the most impressive grass roots mobilization" they had ever seen. There are over 370,000 hair salons and 200,000 pet groomers in the US, and they each cut an average of three pounds of natural fiber a day. These two sectors mobilized in less than two weeks to send 750,000 pounds of hair and fur to the Gulf at their own expense in order to help clean the gulf after the BP Horizon well exploded. The owners of 19 huge warehouses right on the water, from the Florida Keys to Louisiana, donated space for the natural fibers to arrive so that communities would be prepared when the oil came in. Over 2,500 alpaca, llama, sheep and buffalo farmers in the US trucked in waste fleece to save the shores. Hanes, Spanx, Hooters waitresses and thousands of individuals donated over half million pairs of nylons which were cut to make tubes (booms) to contain the natural fibers. Ben & Jerry's and their coffee bean partners Green Mountain donated pallets of coffee bean burlap sacks to wrap the booms so they could be dragged over rocks and behind boats.

Thousands of volunteers made mountains of booms at the donated warehouses, at "boom B-Q's" and cut-a-thons all over the world and then shipped them to the Gulf, ready to donate to BP before the oil came ashore. Every town in North America and 30 countries contributed hair cuts, fur clippings, waste fleece and stockings to this natural fiber recycling effort.

*Q: Despite all of this, BP still refused to use the hair booms in the clean-up efforts and blocked your efforts. How come?*

**Lisa Craig Gauthier:** A few people at BP decided that it was preferable for BP to stand by their message that BP's synthetic oil-based booms and dispersants are superior products and the only ones right for the job. Oil companies use oil-based products to clean up oil spills. Period. Parishes and cities in the Gulf are so cash strapped they really couldn't risk not working with BP as long as there was hope BP would pay for the cleanup. After BP left, public works departments, private beach front properties and harbors took some of our donated booms with the promise they would dispose of soiled booms properly in BP-designated hazardous waste landfills in Florida, Alabama and Mississippi. Harbors have little oil spills all the time and use up booms. Our guess is that 15 percent at most of all the booms got used for the BP spill. It would have been so much better to not use dispersants and collect the oil in the water with natural booms before it got to shore. The Corexit dispersants made it impossible to consider composting any toxic-waste-soaked booms.

*Q: Do you have any tips for community folks and grassroots bioremediators wanting to engage in oil spill response?*

**Lisa Craig Gauthier:** Yes! Don't wait for the oil companies. They will not help you. If it is near your water source get it out fast: use homemade hair booms and dispose of them according to hazardous waste policies. Do not wait — as we can show you so many sad examples of places waiting with high cancer rates after 20 years and still soaked in oil.[1]

## Straw or Myceliated Straw

Sometimes regular oil spill clean-up professionals will use straw on a beach or ground to sop up oil. You can take this one step further by using straw that has been inoculated with mushroom mycelium, so that not only does the straw absorb the oil, but the mycelium will work to break it down (as well as increase the absorption capacity of the straw). In a paper entitled "The Petroleum Problem," Paul Stamets stated that straw that has been inoculated with oyster mushroom mycelium floats,

Fig. 9.2: *Straw barrier
in Dauphin Island.*
CREDIT: © DANIEL BELTRA /
GREENPEACE

allowing it to be used as a sorbent/sponge for picking up oil on water,
as well as a filter and floating remediator.[2] Robert Rawson, one of the
founding members of the Amazon Mycorenewal Project and an expert
in bioremediation and oil spills, told me that he prefers using straw as
a sorbent to hair mats, as mushrooms breaks down the straw more
easily. According to Rawson's experience, hair is more difficult to break
down, and hair mats only make sense if you plan on using them several
times to collect oil.[3]

## MycoBooms™

Paul Stamets is currently experimenting with straw colonized with
oyster mushroom (*Pleurotus ostreatus*) mycelium encased in hemp
tubes. Unlike other booms, these booms are completely biodegradable. Oyster mushroom mycelium in these booms emits enzymes
which break down oil, and the booms support oil-eating bacterial

Fig. 9.3: *Mycobooms™ are an invention by Paul Stamets which corral and degrade oil spills using oyster mushroom mycelium colonized on grasses — in this case, wheat straw.* CREDIT: PAUL STAMETS

communities as they age. MycoBooms™ are still in the experimentation phase, but definitely look promising. Stamets has also found a species of oyster that can tolerate salt water (as not all mushrooms can), which is key for marine oil spill response.

## Peat Moss

As many gardeners know, peat moss is highly absorbent. It is however a fairly non-renewable resource that can be quite environmentally damaging to extract. But if in a pinch you can either raid your local garden centers or get them to donate it, it could be a good option for sopping up oil.

## Coconut Coir

Coconut coir is a byproduct obtained from the husks of coconuts as those husks are processed to make fiber ropes and other articles. If you live in a place where there are no coconuts, you can either order it or find it at your local garden center, as it is becoming a popular, super absorbent material to add to potting mixes. The coconut coir readily absorbs oil and remains in a form that can be easily recovered by scooping or raking. According to some sources, once recovery is complete, the oil-saturated coconut coir can be subjected to squeezing by pressure to liberate the oil collected and then reused for subsequent clean-up efforts. Unlike peat moss, coir is a renewable resource, cheap and easy to find in tropical climates.

I asked Mia Rose Maltz from the Amazon Mycorenewal Project how she thought the different sorbents could be effectively mycore-mediated. Her answer was: "Yes, oil-soaked peat moss or coco coir could work well, with a loose weave. Peat is the product of a climax ecosystem, therefore not a very renewable or sustainable resource. I believe that coir has moderately anti-fungal properties, but I believe that coir mixed with straw and sawdust could be an extremely successful substrate for absorbing oil and then growing saprotrophic fungi on the saturated substrate. The cool thing about hair is that it 'adsorbs' oil, not just absorbs it. I like the mix idea with some hair in it, too!"[4]

## Microbial Remediation of Oil Spills

On a land-based oil spill, you could apply the different tools from Chapter 4 in creative ways. Using bacteria to degrade hydrocarbons is fairly common and has been used by professional clean-up folks like Robert Rawson, Anita Burke and others. Rawson, a professional bioremediator, uses a specific bacterial inoculant along with windrow composting to break down oil in large amounts of contaminated soil. See Chapter 4 for more information.

You could try spraying regular and large doses of compost tea (or some professionally-made and purchased (or donated) bacterial inoculant specific to degrading oil) on the oil slick to unleash the power of microorganisms to break it down. Or you could mix tea or inoculants with compost to also give microorganisms a good habitat and some other food sources. Be aware that you could potentially increase the volume of contaminated soil if your remediation is unsuccessful. You could also add in a nitrogen amendment, like kelp or seaweed, or an amendment with phosphorous, like chicken manure or bone meal, to boost the activity of your microorganisms.

When oil contaminates water, oxygenating the water is a good bet, as that helps with stimulating microbial action. So if you have the ability and equipment to aerate and the right circumstance (a smaller and more contained water body), then go for it.

## Mycoremediation of Oil Spills

As discussed in Chapter 6, mushrooms have huge potential to assist in oil spill response. If you can figure out how to effectively cultivate them or find a good source for spent spawn, their use makes most sense in dealing with the oil once it has been sopped up or contained.

Having enough mycelium available to deal with huge amounts of oil is one of the biggest challenges. In most situations, there will likely be more oil spilled than there is mycelium available and ready to deploy. Shipping large volumes of mycelium is expensive and time consuming. The more local folks you have in your region who are cultivating mushrooms, the more mycelium you will have on hand if a crisis ever arises. If we create more sources of mycelium before the time comes, we can respond in a meaningful fashion with this powerful remediator. Furthermore, many medicinal mushrooms grown (like reishi, oyster, turkey tail and shiitake) are not only powerful healers of the land, but also help our bodies be resilient and recover from toxic exposure, as well as fight off cancer. So having fresh sources of these mushrooms for both before, during and after disaster consumption is definitely a bonus!

It is also important to consider what medium the fungi will be remediating (e.g. saltwater, freshwater or land) as that will determine what mushroom strains or species can be used. Furthermore, we must remember that different oil types will respond differently to mycoremedition. Mycelium more readily degrades hydrocarbons with lower molecular weight (three, four and five-ring, lighter crudes) than heavier-weight hydrocarbons. However, the heavier-weight hydrocarbons can be reduced via mycelial enzymes into lighter-weight hydrocarbons, allowing then for a lengthier, staged breakdown through subsequent mycelial treatments.

Anita Burke has also found mycoremediation to be successful on the ground for oil contamination. "I used mycoremediation on a site in a suburb of Seattle. There was this lot at a very busy intersection right in the middle of town. There were these mountains of contaminated dirt and inside of that was a giant pond," recounted Burke. "I decided that I was going to find Paul Stamets and get some mushrooms from him. And that's what I did. I basically had a mushroom farm in the middle of this town. The earth turned white with all the mushrooms growing on it. Totally cleaned it up. And it did it faster than anything else I would have done."

According to Burke: "Mycoremediation is very important, especially when you have industrial locations where people aren't there very often and soil needs to just sit and get healed anyway. The turnaround times for mycoremediation are far better than some of the other things they try to use, like vapor extraction systems. I think that mycoremediation has a lot of potential for on land spills and contamination."[6]

## *Cosco Busan* San Francisco Bay Spill: Grassroots Mycoremediation Efforts Blocked by Government Regulators

On November 7, 2007, the *Cosco Busan* container ship struck the San Francisco-Oakland Bay Bridge and spilled around 58,000 gallons of bunker fuel (heavy fuel oil) into the San Francisco Bay. The spill spread rapidly, affecting a large area of the California coast. Around 200 miles was oiled. Fifty public beaches were closed, and 6,849 birds, as well as fish and sea mammals, were killed.

Some very passionate folks attempted to do some grassroots mycoremediation on this oil spill much to the chagrin of the company and the government. Several hundred volunteers, led by Lisa Craig Gauthier from Matter of Trust, used human hair mats to soak up the oil from the sands of Ocean Beach, just south of the Golden Gate Bridge. Matter of Trust had been stockpiling these natural hair mats, which they felt worked just as well as the conventional polypropylene mats. Dressed in hazmat suits, volunteers wrung the hair mats out into large dumpsters, collecting several thousand gallons of oil. A few days after the spill, Gauthier asked Paul Stamets to help set up a mycore-mediation demonstration. Stamets agreed and donated hundreds of pounds of oyster mushroom mycelium. Unfortunately, authorities in-tervened, confiscating the oiled hair mats under the guise that they needed to be held as potential evidence for legal proceedings, as well as citing liability and safety issues.

Blocked but undeterred, folks decide to go ahead experimenting; with so much mycelium on hand, they substituted used motor oil instead. Relocating to a small plot of federal land near the Golden Gate Bridge, the team stacked hay bales to create an enclosure with eight chambers lined with a thick waterproof tarp or pond liner. In some of the chambers they mixed oyster mushroom mycelium on various substrates (straw, sawdust and grain) with the motor oil. By mid-January, mushrooms had sprouted.

Grassroots mycoremediator Mia Rose Maltz participated in the experiment. "There were some issues using pond liner in those bays because it did not allow for drainage. Also, the hair mats were difficult to break down/biodegrade because they were very tightly woven," she recounted. "But from the sampling at the end of the experiment, we found that the mixture of spent sawdust spawn, spent straw spawn and rich compost had significantly reduced PAH concentration, as well as all petroleum hydrocarbons present in baseline samples."[7] According to Lisa Craig Gauthier, the experiment had mixed results. "It took us two years, and the mushrooms were only a small part of the cleanup. We moved on to bacteria, thermophylic composting and vermiculture, which eventually after backbreaking months did remediate. And that was just a small spill."[8]

When asked what sort of infrastructure a community would need to have in place in order to be able to respond to an oil spill with mycoremediation, Maltz suggested that the following things were important:

- trained people with a skill set
- prolific spawn resources (e.g. mushroom farm) that could donate spent and virgin spawn
- maps of substrate available by the truckload
- water
- nitrogen (e.g. kelp)
- a community space to teach hands-on techniques to people

According to Maltz, "A little bit of nitrogen will increase the effectiveness of oil spill remediation. Add a little bit of kelp fertilizer, solubilized seaweed or very dilute urine as a nitrogen source, to help keep the mycoremediation going. Often nitrogen is a limiting factor in petroleum biodegradation and bioremediation."

Furthermore, Maltz added: "When I helped with the San Francisco spill, I was trained as an SF Disaster Worker Volunteer. After the *Cosco Busan* spill, on November 27, 2007, the San Francisco City Council unanimously adopted Resolution 123–07 with consideration for studying mushroom remediation and other healthy, inexpensive alternatives to current standard practices for oil spill training and response. Therefore, knowledge of those municipal objectives and programs are helpful — so people could plug in as needed, continue to refine protocols for oil spill response and document mycological strategies for restoring ecosystem resilience."[9] Maltz is a founding member of the Amazon Mycorenewal Project, where they are experimenting with mycoremediation of oil-contaminated pits left by Chevron Texaco in the Ecuadorian Amazon. For more information on their efforts, see Chapter 6 for the full interview.

---

Paul Stamets believes that we need "a large-scale effort to train 'Mycological Response Teams.' These teams would have the skills and knowledge to be able to react to catastrophic events as well as lead others, and be embedded in communities, especially those closest to risk."[10] I couldn't agree more. If we can get enough people trained in many different bioregions, inoculate the grassroots so to speak, then I believe that mycoremediaton could eventually find its way onto the front lines of oil spills.

## Manual Shore and Beach Cleanup

As a grassroots bioremediator, you can try your hand at laying out natural sorbents and remediating them with mushrooms or composting them down, but there is also the less glamorous work (not that the bioremediation pieces are that glamorous either) of simply cleaning up a beach by hand. That can involve raking through oiled sand, shoveling oil sludge or sifting through sand with a colander to pick out tar balls (sounds a bit ludicrous and overwhelming, but it is a common practice). Manual cleanup can also involve pulling oiled vegetation and placing it in trash bags (to either be bioremediated later or in conventional cleanup, disposed of). To engage in this part of the cleanup, wear protective gear (because you want to avoid inhaling or touching

Fig. 9.4: *Oiled beach in Louisiana.* CREDIT: © DANIEL BELTRA / GREENPEACE

crude). In conventional clean-up operations, shore and beach cleanup are usually where many community members end up volunteering their efforts because the work requires considerable human power and does not require previous experience.

## Caring for Oiled Wildlife

Although this topic doesn't fit into the bioremediation we've talked about before, it is an important activity that community folks and grassroots remediators can engage in. Caring for wildlife helps reduce the damage of the oil spill on an ecosystem or at least attempts to alleviate the short-term suffering of affected wildlife. One of the most heartbreaking and prevalent images that come out of oil spills are of birds, fish, marine and terrestrial mammals covered in oil — some still alive, others completely submerged and dead.

In any large oil spill, wildlife casualties can be high. Usually the number of dead wildlife counted comes from those that wash up on the beach or die later in wildlife recovery stations. However many more die at sea and are never counted, and there is very little certainty as to the long-term survival of those who are recovered and de-oiled, especially if contamination of their habitat and food sources persist. Toxic poisoning and internal damage can appear after oiled wildlife have been released, decreasing their vitality and health to the point where they are unable to meet their basic needs. The combination of contaminated food sources and destroyed habitat (resulting either from the toxic contamination of the spill or the destruction to sensitive nesting and feeding areas during mechanical cleanup) makes life pretty dangerous and uncertain even for hardy survivors.

According to the International Bird Rescue: "When oil sticks to a bird's feathers, it causes them to mat and separate, impairing waterproofing and exposing the animal's sensitive skin to extremes in temperature. This can result in hypothermia, meaning the bird becomes cold, or hyperthermia, which results in overheating. Instinctively, the bird tries to get the oil off its feathers by preening, which results in the animal ingesting the oil and causing severe damage to its internal organs. In this emergency situation, the focus on preening overrides all other natural behaviors; including evading predators and feeding, making the bird vulnerable to secondary health problems such as severe weight loss, anemia and dehydration. Many oil-soaked birds lose their buoyancy and beach themselves in their attempt to escape the cold water. The fortunate ones are taken in by concerned citizens or capture crews."[11] Heavily-furred marine mammals, such as sea otters and seals, are affected in a similar way as oil coats their fur and reduces its insulating abilities, leading to difficulties maintaining body temperature and hypothermia. In the cases of both birds and mammals (terrestrial or aquatic), ingesting oil (either via preening, drinking or eating oiled food) or by it entering the lungs or liver can lead to death.

Under these circumstances, most oiled wildlife will die without human intervention. Since the stress of de-oiling, captivity and subsequent reintroduction into compromised environments may cause more suffering and lead to death anyway, is it worthwhile to recover and de-oil wildlife after a spill, or is it is kinder to let them die or euthanize them to ease their suffering?

According to Paul Horsman, "There is a classic example in the Persian Gulf when there was a massive oil spill after the 1991 Gulf War. There is a bird there called the Socotra cormorant. Ninety percent of the world's population of Socotra cormorants lived in the Gulf, and there was a genuine fear that the oil spill would lead to their extinction or at least their massive demise. But these birds had a highly successful rate of recovery from cleaning, and no one would have known that had they not set up the clean-up camps. It is true that some species do not recover very well. There is the stress of the oil, the stress of being handled, the stress of being captive. It depends on the degree of oiling too. If you have a pelican that has been severely oiled and has ingested a lot of oil in trying to clean itself, its chances of survival are pretty small. But if you get a bird that has just recently been oiled and you can somehow stop it from ingesting the oil, its chances of recovery are increased. So it's no hard and fast rule, but I would argue that it is well worth making every effort to try to recover as much as possible. In the end, you are in a mess. What are you going to do to try to make it as less of a mess as you can? Sometimes that means getting your hands dirty and helping."[12]

Similar to oil spill response, time is of the essence. Once an animal has been oiled, the first 48 hours of response and care are critical to its survival. If an oil spill happens near you, there are several different ways to assist in providing care to affected wildlife. You may be lucky enough to have a professional oiled wildlife clean-up organization or a nearby wildlife rehabilitation center to take on coordinating and mainly staffing de-oiling and rehabilitation efforts. You may have veterinarians, wildlife biologists and technicians volunteer their time (or have it covered hopefully by the company at fault or the government). If that is the case, then these folks will likely set up the appropriate enclosures, wildlife stabilization facilities, cleaning stations and rehabilitation pens, or transport living wildlife to existing recovery facilities. Often in oil spills, governments ask that citizens refer this work to wildlife or clean-up professionals, both for your own safety in handling potentially toxic and dangerous wildlife, and to ensure the best care for the wildlife. Wildlife capture, handling, stabilization, de-oiling, rehabilitation and release require training and expertise. Without the proper know-how, well-intentioned but inexperienced help can do more harm than good.

If professional care is available, the pros will likely need volunteers. Some oiled wildlife care and response groups offer basic first responder courses that provide you with some basic skills so that if an oil spill occurs, you can plug into their efforts and assist with capture, transport, stabilization, and basic care of oiled wildlife. I highly recommend taking such a course. In the absence of an oiled wildlife first responder course, volunteering at a local wildlife rehab centre can at least familiarize you with wildlife care and give you some relevant training and experience. If you do not have any training or experience, there are many other ways to assist in a professional wildlife de-oiling response. Crowd control, cleaning cages, prepping food, doing laundry, and collecting donations of supplies are just a few of the many important jobs that will need doing. Sometimes a professional response may not available to you or your community, and this may be the case if you are in a more remote location. Or professional help may be slow in coming. In cases like these, it's up to you and other volunteers to take on this vital work. For those hopefully rare instances when no professional wildlife clean-up effort is forthcoming and grassroots folks have to scramble to respond, below are some useful considerations.

### Facilities and Materials

You will want to have a space, depending on the size of the spill, that is large enough to accommodate up to hundreds of mammals and birds in big spill, less in a small spill. This space must have good ventilation as well as some heated spaces for drying off animals/birds and providing them with a space to warm up. The space should also have water access, and places for hoses and sinks. The cleaning area should have lots of airflow for the health of the workers/volunteers and wildlife.

You will want to have the following donated or purchased:

- hoses
- tubs
- bins
- kiddie pools
- pens and cages (netted-bottom cages with soft sides are great)
- heaters, heat lamps, hot water bottles or pet dryers
- towels, sheets and blankets
- food for wildlife
- transport boxes for crews or citizens bringing in wildlife

· Get the proper gear for yourself and your crew:

- ❧ oil resistant long rubber gloves
- ❧ Tyvek® suit or rain gear
- ❧ face shield (depending on chemicals used and extent of oiling, a respirator may also be a good idea)
- ❧ eye goggles

### Professional Guidance and Volunteers.

It would be best if you had wildlife experts, your provincial/state or federal fish and wildlife agencies, wildlife rehabilitation centers, universities, zoos, aquariums and veterinarians leading the work. But if for some reason they are unavailable to you, then at least try to get them to send you information, protocols and consult with you over the phone. There are several organizations (such as the Oiled Wildlife Care Network or International Bird Rescue) that seem to either set up wildlife de-oiling operations at spills or have lots of knowledge on the topic. Call or e-mail them for more information. Or if you live in a place where you think there could be a major oil spill, consider setting up a yearly training for interested community volunteers to start embedding skills and knowledge locally.

### Capture and Transport of Oiled Wildlife.

When volunteers or responders find an oiled bird or mammal, they need to carefully handle it; most wildlife is not used to human contact and therefore finds handling, capture and transport very stressful. Keeping people and dogs away from oiled wildlife is important. Work in pairs to capture oiled wildlife and do so as calmly and carefully as possible. Place the oiled wildlife in a medium-large box with a towel or soft padding at the bottom. The box should have many air holes, and these holes need to be large enough for adequate airflow.

When transporting the bird or mammal, try to speak softly and avoid sudden movements, loud noises and music. Keep them in a warm and quiet space until they can be delivered to the cleaningnfacility, and minimize contact and handling.

It is also really important that any dead oiled wildlife be collected and disposed of in a safe way; this removes them from the food chain of other animals and avoids increased contamination.

### Stabilization of Oiled Wildlife.

According to the International Bird Rescue "Many oiled birds die because well-meaning people, anxious to remove oil from feathers, wash them immediately, resulting in extreme stress." [13] Many types of animals, and especially oiled birds, really benefit from some rest, hydration, clean food and medical treatment prior to cleaning, so that they may regain some of their strength and increase their chances for recovery. Some birds and mammals will be stronger than others. But most of them will be dehydrated, stressed, traumatized, hungry and cold. Some rehabilitation operations, such as the Oiled Wildlife Care Network, allow impacted wildlife to stabilize (hydrating and feeding them) for at least 48 hours for birds (mammals may require a shorter stabilization period) before beginning the washing process.

For wildlife intake, assessment and stabilization, it would be highly advisable to have veterinary and wildlife professionals present. Handling of animals must be minimal. Any oil in and near their eyes, mouth and nostrils should also be flushed out with saline solution. During this time, it is critical that the birds and mammals are able to regain a healthy body temperature as soon as possible. Once warm

Fig. 9.5:
*Oiled brown*
*pelicans in*
*Louisiana.*
Credit: © Daniel Beltra
/ Greenpeace

and dry, birds should stop obsessively preening, thereby decreasing the amount of oil they ingest in the process. They should be left to rest and be hydrated and fed in order to regain enough strength to undergo the cleaning process.

### De-Oiling the Wildlife

Washing and rinsing oiled wildlife is a very challenging job and one that requires training and skill. As it is very stressful for the wildlife, you want to get it done as quickly and as thoroughly as possible. Birds, reptiles, amphibians and different types of mammals all handle de-oiling differently. Consult a professional about the best way to proceed depending on what you are dealing with. There are also some great online resources with more information on different protocols.

### Wash

Several bins, tubs or kiddie pools can be set up and filled with water and some form of dish soap in a low concentration. Dawn dish soap, especially the blue kind, is used by most recovery operations. It has the ability to get rid of most of the oils, is effective at low concentrations, rinses quickly from feathers and fur, is cheap and accessible and is supposedly non-irritating to the eyes and skin.

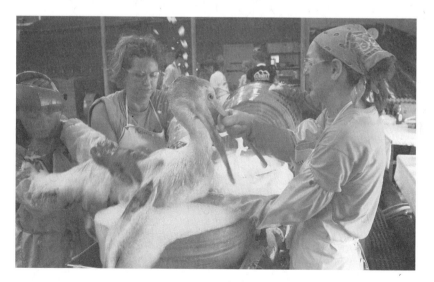

Fig. 9.6: *Oiled white pelican in Louisiana.* CREDIT: © DANIEL BELTRA / GREENPEACE

Depending on the size of the mammal or bird, you will likely need to have one or two volunteers hold it, while someone else does the washing and rinsing. Volunteers and cleaners should wear protective gear, such as rubber gloves, rain gear and likely a face shield and goggles (as the dish washing liquid may be hard on your skin, and the oil will be splashing and misting up with the water and this can be toxic to breathe in and to touch). The gloves can also be a bit of a barrier if stressed out wildlife decides to bite or nip!

## Rinse

Once a thorough washing is done, the bird or mammal is moved to a rinsing station (multiple tubs full of clean water ). The rinsing stage is just as important as the washing — all detergent needs to be rinsed from the bird/mammal. Using a hose, any traces of the detergent and remaining oil are rinsed off the bird or mammal.

With both washing and rinsing, consider where and how you will dispose of the used oily water. Depending on the severity of oiling, avoid pouring it outside or down the drain if you can, as that could lead to contamination of the surrounding environment.

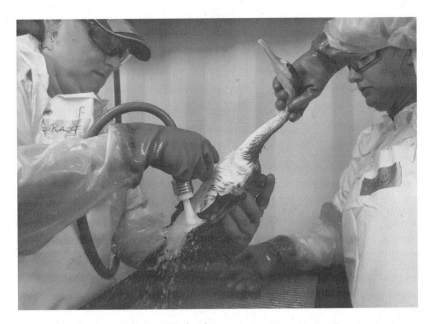

Fig. 9.7: *Oiled cormorant in New Zealand.* Credit: © Dean Sewell / Oculi / Greenpeace

## Dry

When all detergent and oil is removed, the bird or mammal is placed in a pen where it can be dried. Some cleaning operations have commercial pet grooming driers that help with this. If you do not, placing this pen near warm air or nearby heater would be best. Towel dry if the animal or bird will allow you. Make sure the bird or mammal has access to food and water at this point. If it is still too sick or stressed to eat for itself, oit may need to be fed it by hand or tube.

At this point, cleaned birds will start to preen their feathers back into place — this preening allows their feathers to be reoriented into the tight overlapping panel that gives their body a waterproof seal.

## Conditioning and Rehab

Once dry, the mammal or bird can be moved to either appropriate housing or an outdoor pool, where it can continue to recover, groom and regain its waterproofing, while being fed. Human contact should be minimal to avoid stressing the wildlife or allowing it to become to habituated to humans.

## Release

Finally, once the animal is in better health and has regained its waterproofing, it can be released. It is really hard to judge what is the right time for release of individual animals, and how and where to release them unless you are a veterinarian or wildlife specialist, so try to defer to a professional on this.

In the end, I'm sad to say that a large amount of de-oiled wildlife still perish in the days, weeks and months that follow their cleaning. But, as long as you feel that you are not causing more harm to the/ bird/mammal, your efforts and assistance can make all the difference and really help individual wildlife and their communities cope with the disaster.

## Organizing for Community Disaster Response

Responding to an oil spill is responding to a hazardous waste clean-up situation. In disastrous spills, what can your community safely do? Just as knowledge is power when it comes to better understanding the nature of an oil spill and the conventional methods used to handle it, let's explore a few preventative measures to minimize the impact on yourself, your community and the Earth in the event of a big spill. Many of these measures also apply for smaller spills.

### Demand that dispersants and in situ *burning not be used unless it is absolutely necessary.*

These clean-up methods are controversial and can create a potentially dangerous environmental and health situation for your community. There are times when using a dispersant may be warranted, but it is rare. Get an independent, non-company, non-governmental expert opinion. Not using dispersants, especially in huge quantities, could reduce a huge host of toxic impacts, for both people and wildlife. Some studies have found that the chemicals used in dispersants can be more toxic and fatal than the crude itself.[16]

If dispersants end up being used, which is likely since they are favored by industry, make sure that dispersants are not being sprayed around you, upwind of you or on you (this has happened accidentally in clean-up operations). Make sure they are not being sprayed in conditions that would disperse them towards other clean-up crews or vulnerable populations. Same is true of *in situ* burning; if a company is doing this, make sure to evacuate folks from downwind communities or clean-up operations.

In the first 48 hours, a lot of the volatile compounds in oil evaporate, causing some pretty intense air pollution that should be avoided as much as possible for health reasons. People generally notice the smell, or feel effects such as dizziness, headaches and nausea.[17] The company and the government will not necessarily prompt you to leave the area or to make arrangements for more sensitive communities members, so it is up to you to make the call. Elders, children, pregnant women, any one with autoimmune disorders, cardiac or respiratory illnesses should be evacuated/relocated away from spill and definitely not stay downwind in the week following the spill, especially if dispersants or *in situ* burning are used.

If dispersants are used, understand that no matter what government or companies say, it is likely not safe to swim in the water or eat shellfish or fish from those waters for some time (unless you get them independently tested and find them to be clean). Find out which dispersant is being used, as different dispersants have different health impacts and degrees of toxicity. Finally, if dispersants are used, take preventative health measures to support your body in handling them better (see Chapter 11).

## Demand protective gear, hazardous materials training, full disclosure of chemicals being used and their health impacts.

In a crisis and in the rush to respond, especially when efforts do not go as planned, precautions get skipped and people are asked to step up and take great risks by putting themselves in incredibly toxic situations. If you are working with the company or the government to respond to the spill, make sure the company provides you with the proper hazardous materials training (which is considered mandatory but sometimes skipped) and proper safety gear and personal equipment for the work you are doing. Taking the proper precautions, using the appropriate gear, as well as taking time to rest and detox will go a long way in decreasing the potential of more serious and fatal health conditions later, as well as overexposure to chemicals.

I can't stress this point enough: the company will not always provide you with the proper gear, and you can almost always expect them to downplay the risk of the impacts of the chemicals you are being exposed to. This happened during the *Exxon Valdez* spill, and it happened over 20 years later for fisher folks, community volunteers and clean-up workers at the BP Gulf Horizon spill, and it has destroyed people's lives. Maybe it's because we don't live in a precautionary culture and companies are just woefully unprepared with the gear to provide to folks in that scale of a disaster. Or maybe they do not want to raise public alarm and panic over toxic impacts, and outfitting clean-up workers and volunteers in full protective gear would arouse suspicions about the clean-up methods being used and the risks associated with the industry. Maybe it's because the corporate culture sees people, like the land, as dispensable, especially frontline communities. It's hard to tell, but regardless, the proof is in the toxic pudding.

Do your research, figure out what you need to have to be safe and fight hard to make sure you and your fellow responders have it. This also applies for if you are organizing your own response: Suit Up!

## Personal Protective Gear for Oil Spills

Remember that though having the gear listed below will help you in many ways, there are certain factors in an oil spill (e.g. weather, humidity, chemicals eating away at your gear) that can make this gear less effective, so you are never 100 percent protected.

### Respirator

People can get really sick by inhaling (and sometimes ingesting via droplets and oil getting on their face) toxic chemicals, and it's one of the surest ways to end up with some really rough health impacts. Respirators also seem to be the thing that companies never have enough of, and they often claim that respirators aren't necessary. From what I can gather, you most definitely want one, especially in the first several days of a spill, where dispersants may be getting sprayed and the more volatile parts of the crude are evaporating, and definitely in cases of wind or rain that make things more mobile. Remember that under certain circumstances respirators don't always function well. First of all, they need to properly fit you in order to protect you. Sometimes (and I read about this often in the *Exxon Valdez* spill) they can become clogged up. There are different types of respirators and different filters for different activities and contaminants — do your research and figure out which one you will need and how to use it effectively.

### Tyvek® Suit, Tychem® Suit or Full Rain Gear and Tape

You want to protect as much of your body as possible from exposure to chemicals. Those pictures of folks covered in oil? You really don't want that to be you! Protective gear companies offer different body protection suits (like Tyvek® or Tychem®), which can be good at keeping a lot of chemicals off the skin, but there are some problems with them. They can rip quite easily. Some online descriptions say that some suits are good for at least 30 minutes of protection against a range of chemicals — unfortunately oil spill responders often find themselves out for more than 30 minutes, so I'm not sure how that all pans out.[18] In some oil spill situations, volunteers have been told to wear thick rubber rain gear (not Gore-Tex® though), ☞

however I've read stories of some pretty gnarly chemicals eating through rain gear. But any outerwear is still better barrier than regular clothes, and it's especially good for wet and rainy conditions. Wearing protective clothing can make you hot and this was a problem in the Gulf of Mexico, where folks often got overheated and dehydrated pretty quickly. Outerwear can also be really uncomfortable, however it is necessary you wear it. Both suits and rain gear need to be taped shut at the boots and hands to keep oil from getting in.

## Long, Thick, Impermeable Gloves

Pretty self-explanatory. You will be touching crude and other chemicals, and you don't want that to get on your hands (and then into your mouth).

## Goggles or Face Shield

Cheap and easy to find. Protect your eyes with goggles, or protect your whole face with a face shield. The face shield could be good if you are in a situation that involves a lot of splash and you find things getting on your face, into your nose and such and you don't have a respirator handy. A shield won't keep you from inhaling toxic fumes, but it is a barrier to chemicals from physically splashing onto your face and you ingesting them.

## Hard Hat or Rain Hat

Hair absorbs oil. Cover your head, with either a hard hat or some form of waterproof hat, unless your protective suit already has a hood.

## Boots

You don't want to wear sandals for this work. Wear close toed, waterproof, sturdy shoes or boots. Rubber boots are good. In some cases, you will be working on really slippery terrain, so be ready for that.[19]

Consider what you do with this gear at the end of the day. You'll need to either dispose of it or clean it off. When you take it off, take extra care to not get anything on yourself if you can. When it comes to cleaning it, do not bring it into your home with you and only clean it somewhere ☞

with good ventilation. I don't know too much about this piece, except to be aware that your protective gear itself becomes a source of contamination at some point, and you need to watch for that and treat it accordingly. Make sure to also engage in other decontamination practices, such as taking a shower after you are done working on a spill.

### Organize a community spill safety response team and/or community monitoring program.

When oil spills hit, local folks are often not involved in decision-making about the cleanup and generally lack access to good information about how events are unfolding. With big spills, companies often need to hire community volunteers and have to train them before the cleanup effort can begin, losing precious response time. Though most folks seem to mobilize most naturally during a crisis, the proper time for planning and training in spill response is not after an oil spill has already occurred — but before.

If you are living in a community that is at risk of a spill (I would argue that most of our communities, lands and waters face the risk of an oil spill, though some locations face a considerably higher risk than others due to proximity to oil extraction, processing and export infrastructure), it would be advantageous to organize a group or council of community members who can act as a well-informed advocacy and advisory group on oil spill response. This group can also communicate with or pressure companies and key government agencies to disclose the spill response plan so that it can be scrutinized by independent sources and likely beefed up to ensure better response resources and procedures.

Community advocates can pressure a company to make sure that more oil response resources (such as booms, sorbents, boats and skimmers) and protective gear are kept locally, and in a sufficient amount for a strong and quick response. There are many experts, responders and knowledgeable community members in different regions who have lived through past oil spills that can assist your community in knowing what to ask for and what to push farther on when dealing with companies and government.

Training could be sought out on oil spill response, as well as on grassroots bioremediation methods in the event of a spill. Push the company to see if they will cover such training; you can either play it off some greenwashing initiative they have or you can organize and pressure them until they acquiesce.

You should not expect a company to notify you of spills. Sometimes they try to cover it up, and often they underplay it. Many companies have shown a habit of no notification and then slow response and sometimes even no response to spills. It can fall on citizens to report these spills and to convince the government and media of the scale of the problem. If you live in a community where the company has already proven itself untrustworthy and you know you cannot rely on them to either properly monitor or report spills, then organize a local community watch group to keep an eye out for signs of spills. This may mean having folks walk the length of a pipeline, or riding, boating or paddling up waterways to make sure there are no signs of oil sheens.

Work with folks in your community or surrounding communities who could do this fairly easily, such as fisher folks, guides, outfitters, hunters and trappers, helicopter and bush pilots. If you dealing with the threat of an offshore spill and happen to have connections with anyone who has a small plane or helicopter, get them to be your eyes in the air, as flyovers have often revealed what some companies would like to conceal. You can also put an ad in your local paper or posters up around town and the operations site with an anonymous email and phone number for whistleblowing workers to report spills as they see them happen on the ground.

### Create a community spill response or recovery plan.

If the company has a spill plan, maybe your community (if you are living near a pipeline, port, offshore rigs or tanker-travelled coast) should have one too? Coming up with a community spill response or recovery plan is a great way to gather more information about what resources you have at your disposal. Figure out what expertise exists in your community on these matters and also explore how citizen efforts could interface with company and government clean-up efforts. Creating your own plan would also allow your community to have something to fall back on if disaster strikes and time is of the essence. The time it takes and the knowledge required to come up with such a plan will result in some community members increasing their expertise, training on oil spill cleanup and therefore becoming able to share newfound understandings and resources with the larger community if needed.

If you live on a coast or shore, groundwork could be done to map out which parts of that shoreline are most sensitive to oil spill impacts and therefore need to be identified for quick response in the event of an oil spill. Engaging the company or government on your community spill response plan could be advantageous, at least in the beginning, as it would also allow you to better understand their perspectives, tap into some of their field experience and knowledge, increase functional collaboration if a spill happens and potentially build crucial relationships that would hopefully result in a higher chance of key officials or employees feeling responsible and accountable to the local community and better syncing up on the ground when the time comes.

### Build the infrastructure for a grassroots bioremediation response.

Preexisting training and infrastructure are needed in order to be able to effectively respond to a spill. Invest in and encourage folks from your community to attend mycoremediation and mushroom cultivation workshops and training. Have folks seek training in microbial remediation oil spill cleanup or de-oiling animals. Get acquainted with the different safety protocols and gear you will need. Start up a mushroom cultivation operation or farm (or have a good relationship with one), thereby having mushroom spawn and spent inoculated substrate in large quantities locally that would allow for you to respond in that way to an oil spill. Also build and stockpile things like natural sorbents and booms so that you have them on hand. Have several high-volume compost tea brewers available for use. You may also need access to boats, hoses, places to dispose of oiled materials, rakes and shovels.

### Prepare community healers.

Earth care needs to also involve people care. Reach out to local health practitioners and healers to provide them with information on the health impacts of oil spills and symptoms associated with its different chemical exposures so that these healers are able to properly diagnose, provide meaningful treatment and advocate for responders and community members in the event of a spill. Work with herbalists and other healers to come up with potential detoxification treatments for responders and impacted communities, as well as supplements people can take to boost their defenses and abilities to flush out harmful chemicals as they do their work. See Chapter 11 for tips and ideas for self-care in toxic situations.

### Organize, mobilize and resist pipelines, tankers, offshore exploration and tar sands.

The best medicine is preventative medicine. If you don't want a spill to happen in the first place, you need to remove the infrastructure and practices that lead to it. Organize and mobilize to resist, stop and shut down irresponsible projects that cause spills.

## Grassroots Bioremediation Requires Grassroots Resistance

I started writing this chapter in February 2012. Today, as I finish it off five months later, two new oil spills have unleashed their destructive force in the province I grew up in — Alberta, Canada. It is a province infamous as ground zero of the dirty tar sands mega project. I wonder how many more oil spills will happen before *Earth Repair* is even published. Since 2006, Alberta's pipelines have spilled the equivalent of almost 7.5 million gallons of oil.[20] And this figure does not include the toxic seeps from the tar sands tailings ponds that poison downstream watersheds and communities or the spills and leaks that go unreported across that province.

For the last several years of my life, I've been living on the beautiful and wild coast of British Columbia (BC), in unceded Coast Salish territories. If Enbridge and the Canadian government get their wish, they will punch the 730-mile Northern Gateway pipeline through from the Alberta tar sands to the cold coastal waters of Canada's west coast near Kitimat, BC. Between 1999 and 2010, Enbridge pipelines were responsible for 804 recorded spills that released approximately 6.8 million gallons of crude into the environment, an amount equivalent to half of the oil spilled by the *Exxon Valdez*.[21] Enbridge's proposed Northern Gateway pipeline project would include two pipelines:

one flowing west and carrying a daily average of 525,000 barrels of oil sands crude; the other flowing east and carrying a daily average of 193,000 barrels of condensate (toxic hydrocarbons used to thin crude for pipeline transport).

The proposed route of the pipeline would cross more than 785 life-giving rivers. It would cross through the headwaters of the mighty Skeena, Fraser and MacKenzie watersheds, as well as two mountain passes.[22] It would cut through the traditional territories of over 50 First Nations, many of whom view the pipeline proposal and its federal review process as a direct violation of their laws, traditions, values, livelihoods and inherent rights as Indigenous Peoples under international law.

At Kitimat, oil would be loaded onto supertankers. An average of 220 tankers would navigate a challenging route through inner coastal waters yearly, sometimes in bad weather, with limited visibility, and having to make sharp turns and passing through narrow channels. The Hecate Strait, which many of the tankers would need to cross to get to open ocean, is rated by Environment Canada as the fourth most dangerous body of water in the world and has recorded 85-foot waves.[23] Meanwhile, the supertankers proposed for this route would carry up to ten times the volume spilled by the *Exxon Valdez*. If an oil spill from a supertanker were to happen here, the impacts to the coastal environment and surrounding communities would not only be devastating, but oil spill response and cleanup could face many obstacles and delays. If an pipeline leak or spill were to happen over any of the  rivers and streams, or if a landslide were to sever the pipeline along the remote forested and mountain pipeline route, how long would it take before it was discovered and addressed? What would be lost in the meantime?

In July 2010, an Enbridge pipeline burst in rural Michigan, USA, spilling an estimated 843,000 gallons of bitumen into a tributary of the Kalamazoo River. This accident was the first ever recorded major spill into water of diluted bitumen from the Alberta tar sands as well as the largest inland oil spill in Midwest US history. The lighter more volatile chemicals in the diluent and bitumen (such benzene and toluene — both known carcinogens) as well as hydrogen sulfide and oil evaporated creating toxic fumes that impacted 60 percent of the local population and clean-up workers, with many people reporting headaches, dizziness, nausea, coughing and fatigue.[24] In the water, hydrogen sulfide, benzene, polycyclic aromatic hydrocarbons and some crude oil dissolved creating a toxic plume. Meanwhile, the heavier bitumen began to sink to the bottom of the river where it mixed with river sediments, smothering aquatic plants and organisms and making conventional cleanup methods such as skimmers and oil booms useless in recovering the large amounts of submerged oil. Both the US EPA and Enbridge were unprepared to deal with a spill of sunken oil of this scale, prompting Ralph Dolhopf, EPA Incident Commander for the Kalamazoo spill to state that "This was the first time the EPA or anyone has done a submerged cleanup of this magnitude" and that the EPA was "writing the book" on figuring out how to recover large amounts of sunken oil. Several years later and after a very expensive, and many would still say incomplete, cleanup, over 30 miles of the Kalamazoo still remained closed to swimming, boating, fishing and wading, and human and ecological health impacts continue to shake local communities and the river as a result of the continued presence of submerged oil.[25]

Bringing the threat of an oil spill even closer to my home, another tar sands pipeline, the Kinder Morgan Trans Mountain pipeline, has already been constructed. An increasing number of tankers filled with diluted bitumen leave the Vancouver, BC, harbor regularly and are transforming the Burrard Inlet into an export facility for dirty oil. At a recent Kinder Morgan open house in Victoria, I had the chance to speak with a few company representatives about their oil spill response plan. Not a single Kinder Morgan representative could clarify what their plan was for dealing with a bitumen spill. They kept telling me that they would contain the spill and collect the oil with booms and skimmers. But when I pointed out that bitumen can sink in water quickly or become neutrally buoyant, making conventional oil spill containment and collection technology less effective, the representatives acknowledged it as a "potential" issue, said they were working on it and until then they would just address it with their current technology. However, this technology has already been shown by the Enbridge Kalamazoo spill to be ineffective when dealing with a bitumen spill. Not exactly what you want to hear from a company that currently has a pipeline in operation and tankers moving bitumen out of Burrard Inlet, with plans to expand their operations.

So remember this and don't let any slick oil company advertising tell you otherwise: some types of oil, like tar sands bitumen coming through pipelines or being carried by ocean tankers, are very difficult and sometimes impossible to clean up and do not respond well to the more conventional and easy forms of recovery. Hell, even lighter forms of oil in the perfect conditions are still incredibly hard to cleanup. On top of it all, oil like tar sands bitumen is not only more difficult to clean up, but may also be more prone to causing spills and pipeline leaks.[26] Not very heartening news for anyone living near the tar sands or in the current or proposed path of one of its many pipelines, such as the Enbridge Pipeline in Canada or the Keystone XL Pipeline in the USA.

What is the take-home message from all of this? If you want to dedicate yourself to healing the land, engage in preventative medicine and allow your work to extend beyond the physical and into the political. Oppose tar sands, pipelines and tankers. Oppose the offshore drilling and Arctic drilling. And yes in so many ways that is hard to do, especially in our heavily constructed, oil-dependent culture. But oil is a finite resource, and our addiction to it is changing the climate, killing the planet and poisoning frontline communities. We know that when it comes to oil, accidents have happened and they will continue to happen. Unchecked and unopposed, the damage will be too great and devastating to the Earth, the oceans, the rivers and the wild communities who depend on them.

With this in mind, we need to stand in solidarity with the many Indigenous and impacted communities living on the front lines and in the path of these oil pipelines, tar sands mines, tankers and other destructive projects. We need to do whatever is within our individual and collective power to stop these projects from proceeding, from grassroots organizing to mass civil disobedience and direct action. It is easy for many of us standing on the sidelines to say we love the land and the communities we call home. But at a certain point that love needs to be backed up and met with the true weight of our actions — and these actions must ride that fiercely beautiful edge of relevancy and urgency with honesty, foresight and courage. Preventative medicine is the best medicine, and this is where grassroots bioremediation and grassroots resistance merge.

# Nuclear Energy and Remediating Radiation

NUCLEAR RADIATION is the sort of contamination that you never want to have to deal with. The Fukushima Daiichi disaster has proven in present times that no matter what the nuclear industry or governments say, when nuclear plants go down, the situation can go from bad to catastrophic extremely quickly.

Following the nuclear disaster at Chernobyl in 1986, the Soviet government chose long-term evacuation over extensive decontamination; as a result, the plants and animals near Chernobyl inhabit an environment that is both largely devoid of humans and severely contaminated by radioactive fallout. The meltdown in March 2011 of three nuclear reactors at the Fukushima Daiichi in Japan also contaminated large areas of farmland and forests, but lacking land for resettlement and facing public outrage over the accident, the Japanese government has chosen to attempt a decontamination effort of unprecedented scale. Decontamination could involve stripping all the topsoil and vegetation in the area as well as scrubbing down buildings in a battle against contaminants like cesium dust.[1] From what I can gather, the effort is not going too well, and people in Japan continue to be put at risk from high levels of radiation.

Nuclear disasters and radiation don't stay in one spot and their impacts can travel far, as evidenced by the radioactive sea plume that headed towards the west coast of North America and the elevated traces of radiation folks found in seaweed on the west coast post-Fukushima.[2]

It's important to acknowledge that working in a radioactive situation is incredibly dangerous and can be deadly. I'm therefore not recommending you do it, but it is important to know that there may be some things in the grassroots bioremediators tool kit that could be of use if you or your community ever had to deal with radioactive fallout or contamination, whether it be from a nuclear accident, radioactive releases from a nearby reactor, uranium mining, weapons testing or the generation and management of nuclear waste. Protective gear is a must, and limiting your exposure time is also critical. Also, protecting your body with cancer-fighting plants and foods is key, so read Chapter 11 for more information.

## Grassroots Bioremediation Responses

Some studies on bioremediation for radiation contamination were done around the Chernobyl accident in the Ukraine in 1986. In February 1996, Phytotech Inc., an American company, reported that it had developed transgenic strains of sunflowers that could remove as much as 95 percent of toxic contaminants in as little as 24 hours. The sunflowers were planted on a Styrofoam raft at one end of a contaminated pond near Chernobyl and left to see if they would extract radio nucleotides. When tested after 12 days, the cesium concentrations within the sunflower roots were reportedly 8,000 times that of the water, while the strontium concentrations were 2,000 times that of the water.[3] Hemp plants have also been planted around Chernobyl to test their phytoremediation effectiveness, with some favorable reviews coming out.[4] For a list of other plants that have been studied and found to uptake radioactive metals please refer to Figure 1 in Chapter 5.

A phytoremediation project was initiated at the US Argonne National Laboratories where they planted a large amount of shallow-rooted hybrid willows and special deep-rooted hybrid poplars to deal with tritium (radioactive hydrogen) and VOC contamination problem in groundwater. At the Savannah River Site, phytoremediation is being implemented at one of the waste units. Case studies have also been conducted on-site on managing groundwater plumes of tritium and VOCs with pine trees that are native to the area. In 2006, a phytoremediation project was initiated at the Hanford Site, a mostly decommissioned nuclear production complex on the Columbia River in Washington State. This notorious site was part of the Manhattan Project, and plutonium manufactured at the site was used in the first nuclear bomb and in the bomb detonated over Nagasaki, Japan. Decades of weapons manufacturing left behind 53 million gallons of high-level radioactive waste, 25 million cubic feet of solid radioactive waste (mostly buried on-site) and 200 square miles of contaminated groundwater beneath the site, which could/is potentially leaching into the Columbia River. In an attempt to stop strontium from further leaching into the river and to remove it, the phytoremediation project/experiment is using a species of native willow, coyote willow (*Salix exigua* Nutt.), to both phytoremediate and rhizofiltrate radioactive strontium-90 contamination.[5]

Having really vibrant soil ecology seems to be quite helpful when dealing with radiation contamination. I found quite a few  scientific studies that proved the ability of microorganisms to immobilize radioactive elements, and I would recommend taking a bit of extra time to do some research at your local university library or online to see what past and current studies have been done using bioremediation for dealing with radiation.[6]

Earthfort, an agricultural company that specializes in the management of soil biology, has developed a biological soil inoculant (Soil ProVide™) that is showing  promise as a potential aid in the remediation of radioactive soils. Oregon State University is testing the bacteria Earthfort uses in their inoculant to see if it can immobilize radioactive contaminants in soil, specifically cesium-137. Cesium-137 is one of the main contaminants being found around the Fukushima Daiichi nuclear facility in Japan. In preliminary results from a laboratory experiment, the research team found a three- to four-fold reduction in the amount of cesium leached from inoculated soil. Another group of students at OSU is studying cesium uptake by radishes grown in inoculated and non-inoculated soil to see if the bacteria can bind the cesium and keep it from moving into plants.[7]

According to Elaine Ingham, mycorrhyizal fungi have the ability to bind radioactive particles at the roots of plants, keeping them from being draw up and contaminating food crops. "Some research done on the Chernobyl nuclear disaster showed that the only foods that didn't have radioactive materials were those grown in organic gardens that had good levels of mycorrhizal fungi on the root system and strong soil biology," stated Ingham. "None of the carrots, onions, leeks, apples or garlic that were growing in those soils had high levels of strontium and radioactive iodine, those radioactive compounds that were released from Chernobyl."[8]

Some folks have also been promoting biodynamic farming methods as a way to protect the earth from radiation. Supposedly, some European biodynamic farms post-Chernobyl were found to have not received radiation from the fallout. The farmers reported using one of the biodynamic preps, the Barrel Compost, on their lands.[9] The Barrel Compost preparation is special manure that has crushed eggshells and basalt in it.

Among many things, biodynamic farming involves working with biodynamic preparations (similar to homeopathic treatments) made to help bring different cosmic energies and subtler influences into the soil thereby revitalizing it, rebalancing it and protecting it in ways beyond the physical. But it also involves a lot of work that stimulates and builds the soil ecology. The preparations are made in ways that harness and activate microorganisms, which could be a big part of the radio-protective results reported post-Chernobyl.

## Paul Stamets on Mycoremediating Radiation

In my interview with Paul Stamets, I asked him if he thought mushrooms could help heal radiation-contaminated land after disasters like Fukushima. This was his reply:

**Paul Stamets:** With the Fukushima nuclear disaster, those power plants are still leaking radioactivity. Here is an example of one disaster that will take many generations to clean up. In a paper, I proposed to create a Japanese Nuclear Forest Recovery Zone. It's a strategy for making an old growth forest since it will take so long for this radiation to go away. By selectively going in and harvesting the hyperaccumulating radioactive mushrooms, you can, in a stepwise fashion, slowly decontaminate that environment.

This is not a fast solution. It will take hundreds of years, but what is the other alternative? You scrape off all the topsoil. Where you going to put it? The incineration in Japan of radioactive soil caused huge spikes in radiation because the incinerators cannot fully scrub the radioactivity. They have to be rebuilt, but they basically have a whole new generation of incinerators that are incapable of handling that volume. So in absence of any other alternative, I think the best approach is mycorestoration that would go on for hundreds of years and teach our children about the dangers of unforeseen circumstances. To me and to you and to everybody reading this book, we knew about earthquakes, we knew about nuclear power plants, we knew about tsunamis. It doesn't take a rocket scientist to think that all three of them could occur at once. Yet in no manual developed by the nuclear industry was there ever suggested the scenario that occurred in Fukushima, and that shows you how shortsighted the safety planning has been for these nuclear power plants.

Another interesting thing that fungi can offer for dealing with radiation centers upon melanizing fungi. These fungi have the same melanin-like compounds as in your skin that become dark when you are exposed to sunlight. These melanizing fungi have been found to bind up radioactive metals into the least soluble form found in nature. You can't destroy heavy metals; you can just move them, and you can make them less soluble. If you make them less soluble, they tend to not enter into the water and they are immobilized. If you can immobilize radioactivity, you can control the collateral damage it has by keeping it localized in an insoluble form, and this is what these fungi do.[10]

---

When it comes to remediating a nuclear site or dealing with contamination following a nuclear accident, there are few if any solutions that have been found to quickly and effectively address the severe human and ecological impacts from radiation that has escaped into the environment. Cleanup for these sites is at best a very challenging and lengthy affair. There is hope in bioremediation methods, but there are also some significant challenges. Though some plants and mushrooms can extract and bind radiation, their wastes could be considered highly toxic, as they contain radionuclides. You may still be left with some form of radioactive waste that would need special and expensive disposal — and we have yet to find a "safe" place to store nuclear wastes for the long haul. Microorganisms have shown promise at binding radiation in the soil, however there is still much to be learned on how to engage microbial remediation effectively in the field.

As with oil spills, the best medicine is preventative medicine, so organize and mobilize in your communities to challenge the lies that nuclear energy is safe and clean. Mobilize to stop new mines, reactors and waste dumps from being built, and campaign to shut down whatever reactors are still running.

CHAPTER 11

# Self-Care for Grassroots Remediators, Community Members and Disaster Responders

THIS IS A CHAPTER ON SELF-CARE for folks living in communities where the land and water has been contaminated, for grassroots bioremediators wanting to start restoring a contaminated site in their community and for disaster first responders and clean-up volunteers. These folks are often exposed to both acute and chronic doses of the many toxic chemicals and metals that are the legacy of our destructive industrial system. As a result, they are more likely to suffer grave consequences to their health and that of their families and friends. We cannot tend to the health of Earth if we are also in critical condition, and in times of environmental catastrophe we need to be able to support those willing or forced to be on the front lines of destruction.

Earth care and people care go hand in hand and cannot be separated. The planet's life-support systems are our life-support systems. What befalls the Earth befalls us as we are part of that same body and ecology. As we poison the Earth, we also poison ourselves. As we heal the Earth, we heal ourselves. But on the meandering journey from planetary injury to recovery, we may also expose ourselves to chemicals and toxins that could threaten our health.

When I started doing research for this book, I looked into the stories coming out of the Gulf of Mexico, after BP's Gulf Horizon offshore blowout and spill. The stories of devastating health impacts seemed endless and were heartbreaking. Over the last several years, I've heard too many tragic stories from community members and disaster

255

responders who have suffered greatly in the aftermath of their exposure to chemicals during oil spills, nuclear accidents and from living in
contaminated environments.

As we grasp at environmental justice, the social circumstances of
those who are living closest to the land or responding to the damage
directly must be addressed. For this work to be sustainable, we have to
take care of ourselves and of each other. From the healing of our physical exposure to chemicals and metals, to the emotional trauma many
endure from witnessing environmental disasters and ongoing destruction, how do we repair, recover and regenerate our own ecologies so
that we can best serve those of the planet around us?

This chapter collects some of the different ways that medicinal
plants, nourishing foods and alternative therapies could be used to
support folks as they work to heal the land or if they find themselves
in toxic situations. This is just some general information, not a pre-
scription, and it's not complete as I'm sure there are many other herbs
or remedies that could be useful that are not listed here. This information should not be taken as medical advice or instruction, and is not
intended to replace the attention or advice of a health care professional.
If you or someone you know is dealing with the impacts of contamination, I highly recommend you seek out medical attention (conventional
or alternative) immediately, as we are all different and require different
doses as well as different strategies to recover. Some plants react better with our unique systems than others, and different plants can react
poorly with current medications you may be on or existing medical conditions you may have. This information is intended as a starting place
for self-care, offering some empowering options in critical situations.

Fig. 11: Credit: MJ Jessen

## Filling the Gaps: Using Plant Medicine and Food to Prevent and Heal from Chemical and Radiation Exposure

BY LEAH WOLFE, MPH

> *Nature abhors a gap.*
>
> — Aristotle

The human body is like a house. It has plumbing that distributes different kinds of fluids throughout the body. It has a thermostat that regulates body temperature, letting the mind know when it's time to get a coat or when it's time to turn on the internal air conditioning by opening the pores and sweating. The body has the structure of skin and bones to give it form and protect the organs. The structure is built with proteins, minerals and other nutrients. Finally, the body has electrical (nerves), defense (immune) and waste disposal systems.

One of the best ways to protect a house is to build it with strong materials that won't be easily penetrated. Supplying the body with materials to keep it strong is the filling-the-gap approach to maintaining health in an environment that is full of harmful chemicals and radiation.

Supplying the body works because the body operates on a principle called selective uptake. The body will select whatever minerals are available whether they are of the best quality, poor quality or even severely contaminated. For example, if the body needs calcium but it only finds strontium-90, it will use that instead. The structure of strontium-90 is similar enough to calcium that the body will take it up. Strontium-90 is radioactive and when deposited in the bones will release radioactive particles that affect nearby cells.

Harmful chemicals and radiation can get into our bodies, replace vital nutrients and wreak havoc in the tissues at a cellular level. To counter the health effects of chemicals and radiation, plant medicine can be used in three ways: prevention, first aid and recovery.

Although *Earth Repair* is focused on disaster situations, most of us are exposed to toxins like jet fuel, dioxin, radiation and lead on a regular basis. Healing from these exposures is a lifestyle process to maintain good health in a toxic environment. Healthy bodies are more resilient and recover faster in disaster situations. A chart provided at the end of the article lists common toxins and counteracting nutrients.

This information is *not* a guide for self-treatment; it is an introduction of basic concepts in a plant-based approach to healing from chemical or radioactive exposure. Those interested in learning more will find resources at the end of this article. Otherwise, discuss options with informed people you trust such as herbalists, nutritionists or doctors.

## Prevention

The preventive approach to healing the body is to fill the gaps and tighten the structure. This principle is based on the theory that when the body has an adequate supply of building blocks, such as calcium, then it will be less likely to assimilate look-alikes. Strong bodies are more resilient to stress and change, and they are resistant to invaders.

### Building Blocks: Minerals, Vitamins and Other Phytonutrients

Building blocks and substances that strengthen those building blocks allow the body to resist the uptake of harmful chemicals. Plants that have important building blocks are sea vegetables (minerals including iodine, alginate), dandelion (vitamins and minerals), nettle (minerals), red raspberry leaves (minerals), hawthorn berries (bioflavonoids), lemon balm (minerals), milky oat tops (minerals), miso and other fermented foods (minerals and protein), bee pollen (vitamins and protein), whole grains (bioflavonoids, fiber, minerals, protein), dark-colored fruits and vegetables (bioflavonoids, vitamins, chlorophyll), nutritional yeast (protein, vitamins and minerals), garlic and white oak bark (minerals).

### Astringents

Astringents tighten tissues by causing them to contract and close their pores. They may help reduce vulnerability to chemicals and radiation. For example, yarrow and eleuthero show promise in research as plants that protect the body from radiation and electromagnetic waves. Another example is the use of wild cherry bark and kudzu to mitigate symptoms of exposure to solvents during the Gulf oil spill cleanup. Astringents are often included in anticancer remedies. Some astringent plants that have been used to prevent and heal cancer include white oak bark, wild cherry bark, yarrow, Japanese knotweed, dark-colored grapes, ginger and plantain. Japanese knotweed and grapes are a source of the popular anticancer remedy Resveratrol and are also high in bioflavonoids.

## Resins, Gums and Demulcents

Plants that contain sticky or slimy constituents help maintain the integrity of the body's tissues and fluids. Resins and gums tend to draw out and trap unwanted debris in the tissues so they can be more easily excreted. Demulcents are slimy and lubricate the tissues; tissues that are too dry are vulnerable. Some examples of resinous plants are chaparral, tree resins, grindelia, lomatium and balsamroot. Demulcent plants include mullein, marshmallow and slippery elm.

## First Aid

There are three steps to first aid: decontamination, detoxification and emotional first aid.

These steps should be taken immediately when a chemical or radioactive accident occurs.

## Decontamination

Decontamination, by removing clothing and washing affected areas, is the primary first aid approach in exposure to chemicals. This reduces further absorption of the chemicals into the skin and prevents the spread of the chemical. Decontamination should happen in this order:

1. If eyes are affected, use running water or a bottle of water with a squirt top to flush chemicals out of the eyes. Contact lenses must be disposed of and glasses should be washed immediately before being worn again.
2. Contaminated clothing should be removed immediately — if a garment must be pulled over the head, consider cutting it off instead.
3. Decontaminate skin by running cool water over the affected area for at least 15 minutes. Use soap to help break down oils on the skin that could trap the chemical.
4. Contaminated clothing should be placed into double-layered plastic bags and sealed for disposal.

### Detoxification

Detoxification is the second first aid approach to chemical or radiation exposure. To detoxify the body is akin to cleaning the body's plumbing. Flushing the excretory systems helps the body rid itself of toxins, heavy metals, free radicals (hydroxyl (OH) in which at least one electron is unpaired) and other wastes. Detoxification occurs through the release of sweat, tears, urine, breath, vomit and excrement. In an acute situation, all appropriate systems of elimination should be stimulated. For instance, if you know the toxins were inhaled and that the outer tissues (skin, eyes) were also exposed, get into clean air at once and decontaminate. Then recognize that the body has likely started the detox process with coughing, runny nose, tears and red, irritated skin. After the acute phase is over, help the body cleanse by drinking water, eating more lightly for a period of time emphasizing whole foods, fermented foods and plants that stimulate detoxification and heal affected tissues (eyes, skin and lungs).

Plants that stimulate detoxification fall into the following categories: diaphoretics, diuretics, expectorants, emetics, laxatives, bitters, alteratives and resins.

#### DIAPHORETICS

Medicinal plants that promote sweating are called *diaphoretics*. They open the pores to release heat and toxins from the body. They work best if they are taken as a hot tea. Common diaphoretic plants include yarrow, red clover, elder flowers or berries, boneset, mugwort, calendula, chamomile, garlic, horseradish, rosemary, oregano, peppermint, lemon balm, lavender and hyssop. Diaphoretics can be combined with hot baths with Epsom salts or sea salt to increase the release of toxins through the skin.

#### DIURETICS

*Diuretics* promote detoxification by flushing and protecting the renal system (kidneys and bladder). They are often high in nutrients or include astringent qualities — thus providing protection on two levels. Common diuretics include dandelion, nettles, cleavers, goldenrod, uva ursi, chamomile, agrimony, chicory, cornsilk, hyssop, nettles, mugwort, celandine and cranberry.

## EXPECTORANTS

*Expectorants* help remove toxins from the lungs. They may be useful in cases of chemical exposure resulting in respiratory issues, such as solvent exposure. They can be combined with astringents and demulcents to support the entire healing process. Expectorants include wild cherry bark, elecampane, grindelia, mullein, angelica, basil, celandine, coltsfoot, garlic, hyssop, red clover, thyme and yarrow.

## EMETICS

*Emetics* are only useful if a chemical has been swallowed, because at high doses they can cause vomiting. Common emetics are boneset, chaparral, lobelia and vervain.

## LAXATIVES

*Laxatives* are often high in fiber or increase the flow of water to the colon, promoting the evacuation of waste. High fiber plants include flax, psyllium, whole grains, vegetables, legumes, beans and fruit. Other laxative plants include yellow dock, cascara sagrada, senna, chicory, fennel, Oregon grape root and plantain.

## BITTERS

Just the taste of bitter alone is enough to cause stimulation of the digestive system. *Bitters* have long been used to stimulate the appetite by increasing the production of digestive juices. Bitters also help the body remove toxins by causing the gallbladder to increase bile production. Bile breaks down fats and helps the body remove waste and toxins. Bitter plants include mugwort, wormwood, barberry, coptis, dandelion roots and leaves, Oregon grape root, goldenseal, yarrow, boneset, gentian and hops.

## ALTERATIVES

*Alteratives* are often called blood purifiers. Alteratives improve overall health by cleansing the blood, the lymphatic system and mucous membranes. They also help the body flush out toxins and metabolic wastes. These plants include dandelion, yellow dock, burdock, sassafras, spikenard, echinacea, turmeric, chickweed, cleavers, gotu kola, holy basil, prickly ash, celery root and seed, spilanthes and red clover.

## RESINS

Many plants and trees have resins. In herbal medicine, *resins* are used to draw things that are causing poor tissue health out of the body. Resins have a long history of use throughout the world as medicines for drawing out infections in the skin. In the treatment of radiation exposure, plantain is often mentioned in Japanese research. In the Southwest US where many have been affected by US nuclear testing programs, chaparral has been used to draw out radiation and other toxins.

## OTHER DETOXIFIERS

Sour foods and herbs also help detoxify by causing tissues to contract while squeezing out stagnant fluids. Sour foods and herbs include lemon, sumac bark, Japanese knotweed and fermented foods. Combinations of chlorella, cilantro and alpha-lipoic acid are often recommended to help effectively rid the body of heavy metals, but check with a health professional or herbalist to reduce the risk of redistribution rather than elimination of the toxins. Other formulas include red raspberry leaf, chaparral and red clover.

### Emotional First Aid

The most important aspect of helping someone in a stressful situation is to stay calm and spread that calm to other people. One of the best ways to do this is to find something for everyone to do. Most people want to help each other out during emergencies, and those who are experiencing fear, grief or panic often recover faster if they are called on to help. Psychologically, a sense that order is arising from chaos is enough to ground people in the reality of what needs to be done.

To calm or reorient a person, here are a few things you can try.

- Slow the heart and respiratory rate by getting the person to breathe with you. And then slow down further into nice deep breaths until the person feels better.
- If the person consents, rub the center of the back just below the neck. The acupressure points there will calm the person — the gesture is as natural as mothers patting babies on the back.
- Try this grounding exercise for those individuals who seem to be gone from their bodies: remove them from the chaotic scene, and

then ask the person to pay attention to their senses. Start with breathing, then ask them to look around paying attention to what they see, ask them to feel the ground beneath them, ask them to pay attention to what they hear. Avoid evoking sense of smell — it is the most powerful memory trigger and can sometimes worsen a situation.

🖋 Offer herbal remedies: one to three drops of Rescue Remedy or Five Flower Essence, five to ten drops of motherwort tincture (slows heart rate and calms the nervous system), five to ten drops of St. John's wort (helps the person who is overwhelmed and needs courage to get through the situation), five to ten drops yarrow (promotes overall healing and provides protection from chemical and radioactive exposures), inhalations of lavender (calming) or bergamot (uplifting) essential oil.

## Recovery

After first aid and detoxification the crisis period is over, and some measures can be taken to promote healing. After an emergency, it's important to remember the stress and possibly trauma that people experienced. Rest, relaxation, revitalization and resilience are important to solid recovery. Although it may be obvious that rest and relaxation are needed, it isn't always easy to do. It is important to set aside time and space for people to get rest or relax.

### Rest and Relaxation

Plants that soothe and/or rebuild the nervous system can be helpful at this stage. These plants include those listed as high in minerals in the "Building Blocks" section above and a few others that help calm the mind and inspire hope or courage. Some examples of those are St. John's wort, yarrow, chamomile, hyssop, lavender, calendula, agrimony and catnip. Plants that specifically encourage sleep (*soporifics*) include valerian, California poppy, wild lettuce, passionflower and hops.

### Revitalization and Resilience

Working toward revitalization and resilience is a significant part of recovery but also returns us to the recommendations for prevention. During this stage, a person may need to recover from chemical or radioactive exposures or may need to recover from psychological trauma. In either case, the body's protective layers (physiological and energetic) have been compromised and need healing. The plants mentioned in the prevention section will help the body recover by replacing lost or contaminated nutrients and improving the integrity of the body's tissues. Adaptogenic plants also play a role in recovery. As the name suggests, they help the body adapt to changes whether internal or environmental and often have a normalizing effect on the endocrine system. Those plants include eleuthero, devil's club, American ginseng, astragalus, holy basil, hyssop, milky oat tops, lemon, peppermint and rhodiola.

| Contaminant | Common Sources | Toxicity Symptoms | Counteracting Nutrients |
|---|---|---|---|
| aluminum | cans<br>foil<br>antacids<br>pots and pans<br>baking powder<br>some cheeses<br>cooking utensils<br>deodorants<br>plant fertilizers and<br>  gardening additives<br>refined junk foods<br>tap water<br>bleached white flour<br>buffered aspirin<br>coal mining<br>  production and<br>  waste | Alzheimer's disease<br>colitis<br>constipation<br>headaches<br>hyperactivity<br>irritability<br>learning disorders<br>loss of appetite,<br>  energy or hair<br>memory loss<br>neurological<br>  disorders<br>numbness<br>skin ailments<br>thyroid disorders | calcium<br>fiber lecithin<br>magnesium<br>vitamin C<br>zinc |
| arsenic | coal mining production and waste | | |
| cadmium | batteries<br>cigarette smoke<br>coffee<br>gasoline<br>metal pipes<br>plastics | anemia<br>dry skin<br>hair loss<br>headaches<br>immune disorders<br>kidney/liver damage | cabbage family<br>  vegetables<br>calcium<br>copper<br>fiber<br>iron |

Fig. 11.1: *List of contaminants and counteracting nutrients.* CREDIT: LEAH WOLFE

| Contaminant | Common Sources | Toxicity Symptoms | Counteracting Nutrients |
|---|---|---|---|
| cadmium (cont.) | refined foods<br>steel<br>contaminated water<br>coal mining<br>  production and<br>  waste | low blood pressure<br>protein/sugar in<br>  urine | manganese<br>pectin<br>selenium<br>vitamins C<br>  and D<br>zinc |
| carbon monoxide | auto exhaust<br>cigarette smoke<br>smog | anemia<br>angina<br>asthma<br>bronchitis<br>emphysema<br>headaches<br>memory loss<br>respiratory disorders | eleuthero<br>vitamins A, B<br>  complex<br>  and C<br>cysteine<br>bee pollen<br>nutritional<br>yeast |
| chlorine | tap water<br>beverages bottled<br>  with tap water | vitamin deficiencies | vitamins C<br>  and E |
| copper | tap water<br>plumbing<br>electrical fixtures/<br>  wires | mineral<br>  deficiencies,<br>  esp. zinc,<br>  magnesium,<br>  manganese and<br>  molybdenum<br>gastrointestinal<br>  tract irritation<br>mental disorders | manganese<br>molybdenum<br>vitamin C plus<br>  bioflavonoids<br>zinc |
| fluoride | tap water<br>beverages bottled<br>  with tap water<br>fertilizers<br>fluorinated<br>  hydrocarbons<br>mouthwashes<br>toothpaste<br>dental fluoride | abnormal hardening<br>  of teeth and bones<br>accelerated aging<br>cancer<br>brain damage<br>genetic damage<br>immune disorders<br>vitamin deficiencies<br>kidney disorders<br>Down's syndrome<br>mental dysfunctions<br>thyroid damage<br>tumors | calcium<br>magnesium<br>vitamins C<br>  and E |

| Contaminant | Common Sources | Toxicity Symptoms | Counteracting Nutrients |
|---|---|---|---|
| hexavalent chromium | air and water<br>cigarette smoke | cancer<br>gastrointestinal<br>  disorders | vitamin C |
| lead | dyes<br>gasoline<br>insecticides<br>paint<br>plumbing<br>pottery<br>solder<br>scrap metal<br>cigarette smoke<br>textiles<br>coal mining<br>  production and<br>  waste | cramps<br>anemia<br>fatigue<br>headaches<br>insomnia<br>nausea<br>vomiting<br>weakness<br>cancer<br>nerve disorders<br>brain damage | chlorophyll<br>cysteine<br>eleuthero<br>iron<br>legumes and<br>  beans<br>pectin<br>lecithin<br>phosphorus<br>cabbage family<br>  vegetables<br>selenium<br>sodium alginate<br>vitamins A, B1,<br>  B2, B complex,<br>  C, D and E<br>zinc |
| mercury | amalgam dental<br>  fillings<br>fish<br>soil<br>fungicides<br>pesticides<br>some cosmetics<br>film<br>plastics<br>paint | allergies<br>arthritis<br>birth defects<br>cataracts<br>depression<br>dizziness<br>epilepsy<br>fatigue<br>headaches<br>insomnia<br>kidney damage<br>memory loss<br>nervousness<br>paralysis<br>seizures<br>vision loss<br>weakness | cabbage family<br>  vegetables<br>calcium<br>fiber<br>lecithin<br>pectin<br>selenium<br>sodium<br>  alginate<br>vitamins A, B<br>  complex, C<br>  and E<br>cysteine<br>nutritional<br>  yeast |
| nitrates and nitrites | processed meats<br>fertilizers<br>tap water | cancers of bladder,<br>  liver, stomach<br>  and other organs | bee pollen<br>lecithin<br>nutritional yeast |

| Contaminant | Common Sources | Toxicity Symptoms | Counteracting Nutrients |
|---|---|---|---|
| nitrates and nitrites (cont.) | | heart disease high blood pressure | vitamins A, B complex, C and E |
| nitrogen dioxide and ground level ozone | smog | cancer emphysema respiratory disorders | bee pollen eleuthero *Panax ginseng* vitamins A, PABA with B complex, C and E |
| polycyclic aromatic hydrocarbons | smoke from cigarettes, wood, coal, oil and most commercial incense | cancer | calcium pantothenate cysteine iron selenium vitamins A, B1, B2, B complex, C and E |
| endocrine disruptors (alkylphenols, DDT, PCBs, phthalates, parabens, bisphenol A, polybrominated diphenyl ethers, polychlorinated biphenyls, dioxin) | pesticides herbicides fungicides coolants plastics dental materials coal mining production and waste | vitamin and mineral deficiencies endocrine disorders childhood obesity birth defects cancer reproductive organ damage hormone imbalance infertility | bee pollen lecithin vitamins A, B complex and C fermented foods sauna therapy, juice fast and other detoxification methods |
| solvents (n-butanol, n-propanol, xylene, butyl alcohol, diacetone alcohol, n-butylamine, chlorinated hydrocarbons, cresol, diethyl sulphate, | cleaners (household and industrial) oil spill dispersants degreasers paint office products correction fluids, (dry erase markers, adhesives) | memory loss personality changes irritability headache nausea vomiting kidney/liver damage concentration difficulties | |

| Contaminant | Common Sources | Toxicity Symptoms | Counteracting Nutrients |
|---|---|---|---|
| solvents (cont.) dimethyl sulphate, dioxane, ethylenediamine) | glues coal mining production and waste | involuntary movements fatigue seizures weakness depression vertigo dermatitis respiratory disorders gastrointestinal disorders | |
| cesium-137 | radiation | liver/kidney damage reproductive organ damage muscle damage | potassium |
| cobalt-60 | radiation | liver damage reproductive organ damage | vitamin B12 |
| iodine-131 | radiation | thyroid damage reproductive organ damage | iodine |
| plutonium-238 and -239 | radiation | lung damage liver damage reproductive organ damage | iron |
| strontium-90 | radiation | bone damage | calcium |
| sulfur-35 | radiation | skin damage | sulfur chaparral |
| zinc-65 | radiation | bone damage reproductive organ damage | zinc |

## A Note on Ethical Wildcrafting

Many plants are at risk of becoming endangered. Before learning to harvest your own plants, please visit unitedplantsavers.org and find out which plants are most at risk.

## Recommended Reading

All 4 Natural Health.com. *Natural Chelation — Nature's Own Chelation Therapy*. [online]. [cited June 2, 2012]. www.all4naturalhealth. com/natural-chelation.html.

American Cancer Society. *Chelation Therapy*. [online]. [cited November 20, 2012]. www.cancer.org/Treatment/TreatmentsandSideEffects /ComplementaryandAlternativeMedicine/Pharmacologicaland BiologicalTreatment/chelation-therapy.

The Golden Triangle of Natural Chelation. [online]. [cited June 2, 2012]. web.mac.com/medicalveritas/iWeb/Sanctuary%20Cancer%20 Clinic/Natural%20Chelation.html.

Hiroko Furo. "Dietary Practice of Hiroshima/Nagasaki Atomic Bomb Survivors." [online]. [cited November 20, 2012]. yufoundation.org/ furo.pdf.

Steven R. Schechter. *Fighting Radiation and Chemical Pollutants with Foods, Herbs, and Vitamins: Documented Natural Remedies That Boost Your Immunity and Detoxify*, 2nd ed. Vitality, 1992.

US Department of Health and Human Services. "Statement on Chelation Therapy by Dr. Claude Lenfant, Director, National Heart, Lung, and Blood Institute, National Institutes of Health." [online]. [cited November 20, 2012]. hhs.gov/asl/testify/t990310a.html.

*Leah Wolfe is an herbalist with a background in health research and public health. She started the Trillium Center in 2009, an endeavor to improve community health through preventive medicine, education and preparedness. Visit serpentine-project.org to learn more about how we are learning from plants and bringing plants to our communities.*

## Plant Allies for Liver and Endocrine Support, Heavy Metal Detox, Cancer Prevention and Multiple Chemical Sensititivity

I interviewed Guido Mase, who is a clinical herbalist, herbal educator and one of the founders and co-directors of the Vermont Center for Integrative Herbalism. From our interview, here is a summary of some of the helpful information he shared.[1]

A lot of the chemicals and metals folks are exposed to in acute or chronic situations are things like cancer-causing petrochemical derivatives and endocrine-disrupting plastics and pesticides. Depending on the dose, exposure and type, these chemicals and metals can impact our lungs, liver, immune system, reproductive systems, central nervous system, and they can cause cancer and birth defects. When they enter into our bodies, they can disrupt our hormones causing a whole host of problems, as well as overwhelm and damage our liver and its ability to metabolize and detoxify our bodies.

### Hormonal Disruption

Give your body phytoestrogens so that it is less likely to uptake the toxic, endocrine-disrupting forms in the environment (such as PCBs, DDT and dioxin). Some examples of good food/herbs include:

- red clover
- legumes (chickpeas, soybeans)
- licorice
- flax seeds, blueberries, hawthorn

### Liver Support

Exposure to carcinogens or toxins in high enough concentrations can dramatically damage or destroy liver. Without your liver, you cannot protect yourself from your own toxicity or that of the environment around you. Things such as industrial solvents will destroy the liver and cause rapid jaundice. Industrial solvents or PCBs that get into the fats of the body can be very damaging for the liver. Take herbs and foods that provide ongoing daily support to your liver.

*Milk thistle seed* is the shining star of liver detox and liver damage — especially when it comes to dealing with contamination from oils and solvents. This feisty weed is an amazing liver protector and detoxifier, especially in acute situations. Grind the seed up and take two to three tablespoons every three to four hours orally for a week or so, and then follow up with two tablespoons of milk thistle a day for up to six months. It is a very safe plant, with no side effects and can be used for long period of time. It is also very easy to grow.

*Bupleurum falcatum*, also known as Chinese thoroughwax and

sickle-leaf hare's ear, is also a great short-term liver detoxer. It is a Chinese medicine used to treat heat and dampness, congested inflammatory reaction and it is bile stimulating; it squeezes the liver, forcing it to work harder. As *Bupleurum* can be harder on the body than milk thistle and can cause nausea and digestive complaints, it is recommended to use only for two to three weeks. It is pretty easy to grow, is a perennial and member of the parsley family. It is the root that is made into a tincture. Recommended dose is ½ teaspoon of tincture two to three times a day.

Eat as many *bitter plants* such as dandelion root, artichoke leaf, bitter greens (e.g. radicchio, arugula) as you can. A lot bitter plants have bioflavonoids that are key to helping the liver function and decrease inflammation. Interestingly enough, our diets used to involve eating a lot more bitters than we currently do. Nowadays, many of us are not eating enough of bitter plants while at the same time we are exposed to more toxins!

You may need to combine all of this with some *ginger* (with or without fennel) tea to help with nausea and appetite loss, as this can happen in acute poisoning. Ginger is also anti-inflammatory, which will also be really helpful, as toxins cause an inflammatory reaction in the body.

You don't want what is being detoxed (via the bile) to be reabsorbed, so eat lots of *fiber* — tons of fruit and veggies or psyllium/Metamucil supplement to allow the toxins to bind and also be flushed from the system. If someone is already dealing with serious liver damage from chemical poisoning, use the herb celandine (15–30 drops), and it will force bile to come out of the liver. But this person needs to take some fiber 20–30 minutes before doing that to get bile ready to go.

Drink lots of water to support your system and help flush the toxins out.

Acute exposure to a dioxin will give your liver scar tissue, continued inflammation and damage. Once you are removed from the acute toxic exposure situation, you will want to continue taking milk thistle and add things that help with ongoing inflammation. *Turmeric root* is a really good anti-inflammatory, especially the fresh root. Depending on your body mass, you can take two big tablespoons a day, or you can take 6–12 capsules several times a day with food. This will help reduce liver scarring and help your liver heal.

### Dealing with Carcinogens/Cancer-Causing Contaminants

A lot of the chemicals we are exposed to are carcinogenic, and this requires more long-term thinking. You won't get cancer the moment you are exposed, but the initial exposure or chronic exposure can result in cancer later on. There are some ways to protect and heal yourself when dealing with carcinogens.

Use *medicinal mushrooms*. These powerful healers go far in terms of balancing immune function and increasing immune system vigilance to cancer cells, as well as help with multiple chemical sensitivity (MCS), which is a chronic condition which tends to occur when folks have been overexposed to toxic chemicals. If you've been exposed to a carcinogenic substance, you may want to be on medicinal mushrooms for your whole life. What you need is long-term immune vigilance against cancer. Turkey tail is a great mushroom for those who are at risk for cancer (from breast cancer from endocrine disruptors, to liver cancer). Reishi and shiitake mushrooms are not only great for fighting off cancer, but they also support the liver, which is absolutely necessary. Red reishi is a fantastic liver detoxifier and reduces allergies and hypersensitivity, which is why it is good for those who suffer from MCS. Besides fighting off cancer cells, the reishi mushroom decreases the level of irritation and inflammation that our bodies will experience. Reishi can be powdered and prepared in tea or soup. Shiitake mushrooms are used for liver diseases and cancer support and really maintain a vigilant immune system to make sure tumors don't develop. It is important to ingest these mushrooms on a daily basis, and to either wildcraft them or make sure they come from organic sources. As mushrooms are hyperaccumulators of heavy metals, you need to be careful of their source.

Eat *garlic* as well as *brassicas* (broccoli, cabbage, kale and mustard). They all have sulfur-rich compounds that are really important for some core functions of the liver, as well as cancer prevention. Try to take two to three cloves of garlic a day. You need to chop it up and let it sit for a few minutes for compounds to be activated; then eat it up. Alternate with brassicas.

Hopefully, you can weave all of these foods and medicines into a daily cuisine. Treat yourself easily and incorporate these substances into a lifelong process so that you can withstand the daily barrage of modern life. Because you can't hide from everything, make sure your body has what it needs: healing herbs, bitter plants, garlic, brassicas, flavonoids and heavy amounts of medicinal mushrooms.

Also, be careful about aggressively detoxing, as you can do yourself some damage by removing things too quickly. Many chemicals are stored in fat cells, and you may end up inflaming systems in the body if you work too hard and too fast to draw them out. People can have a massive inflammatory reaction to too much detoxification, which is why the medicinal mushrooms, turmeric and anti-inflammatories are so important.

### Heavy Metal Exposure

Ingest *cilantro* in high doses, as cilantro pesto or tincture. In high lead cases, you can cut the lead count down in half. Cilantro works really well in chelating/carrying toxic metal out of the body. Mix cilantro with garlic, pine nuts, oil and Parmesan. Take two to three tablespoons a day of pesto plus two to three teaspoons of cilantro tincture a day, both on an empty stomach. Cilantro gets into the blood and acts as a chelating agent.

During this process you need to drink lots and lots of water and eat lots of bitters.

Eat lots of *garlic*, as chelating will also inflame the liver and garlic helps the liver function. Eat some form of antioxidant-rich, anti-inflammatory, bioflavonoid-rich, vitamin C-rich food such as *parsley*, *blueberries* and *hawthorn*. Or you can take vitamin C.

Take *zinc* during heavy metal chelation — 15–20 mg/day but not above 30–40 mg or zinc has an immune-suppressant effect. Many enzymes in the body use metals as part of their structure, and they grab onto bad metals when they would rather use zinc or copper. Ingested zinc (which body needs and would rather use) will take the place of toxic metals. A great source of zinc is pumpkin seeds, so eat up to a cup a day!

### Dealing with Multiple Chemical Sensitivity (MCS)

For some people, the combined effects of many chemicals or a large exposure of even one chemical due to an industrial accident or spill may cause an illness called multiple chemical sensitivity. If healthy, our bodies have the ability to handle and detox a certain amount of toxic substances, but after a certain point, we can surpass that amount and overwhelm the body's natural ability to clean and protect itself. This toxic overload blows out our capacity to handle any toxic chemicals or elements in our environment and thus results in a hypersensitivity to our environment along with a whole host of health problems. I heard someone compare this condition and our bodies to a wooden barrel. The barrel has a certain capacity to hold water in the same way our bodies have a certain capacity to handle chemicals and their damage. If the barrel is overfilled with too much force it will eventually rupture, resulting in an inability from there on to hold any water at all. Folks who suffer from MCS have often had exposure to some strong toxin in their past.

People with multiple chemical sensitivity can have strong reactions to common chemicals and elements in paint, perfume, cars and building materials where before they never reacted. Signs of multiple chemical sensitivity may include runny nose, itchy eyes, headache, scratchy throat, earache, scalp pain, mental confusion or fog, sleepiness, fast heartbeat, upset stomach, nausea, abdominal cramping, diarrhea and aching joints. Many health professionals tend not believe that MCS is a real illness caused by chemicals; it is often misdiagnosed or dismissed as being caused by emotional distress. MCS is also often mistaken for common allergies, but it is different from allergies as signs appear each time the person is exposed to chemicals, the effects are long-lasting and not seasonal, symptoms appear with less and less exposure, symptoms go away when the triggering chemicals are removed and appear in the presence of different and unrelated substances. Multiple chemical sensitivity has often been reported by oil spill workers in the aftermath of the cleanup and is a serious, life-altering illness.

A lot of people's symptoms improve when toxic chemicals and elements are *removed from their environment*, so keeping a clean, scent-free, chemical-free space is really important for them. Have folks take breaks and leave the city to a cleaner, less toxic surroundings.

Eat *medicinal mushrooms* (like reishi and shiitake), as two of the main systems to support are the liver and immune system. It's great that medicinal mushrooms work on both, as well as reduce hypersensitivity.

Use *adaptogenic herbs* (which help us handle stress and trauma) such as American ginseng, rhodiola and licorice. These herbs reduce the drain on the adrenal and help folks handle stress better. The sheer amount of energy it takes to deal with being in a state of ongoing toxicity wears folks down and causes a lot of stress. This can lead to an adrenal shutdown or deficiency. Rhodiola can be taken when you are feeling sluggish and shutdown; it gives you energy and uplifts your mood. Take licorice in a lower dose (a couple of teaspoons daily), though not if you have blood pressure issues as it can interact with blood pressure medication. With ginseng you can take up to ¼ to ⅓ ounce of root daily, for up to six months or more. It is a very safe, but takes a long time to really activate. Devil's club can be used as well, but it is endangered so be careful with harvesting it. Wild sarsaparilla is also a really gentle, mild, bitter adaptogen.

## Radiation Exposure

In the event of a nuclear power plant accident or meltdown, it is recommended that you ingest a *biologically available source of iodine* — like seaweed — especially in the first several days in order to block radioactive iodine from entering your body.

Rhodiola and *tulsi* (holy basil) both have radioactive protective/antioxidant properties and are believed to stop radioactive damage to DNA. Tulsi also increases the activity of enzymes that repair DNA and destroys mutant cells.

Support anti-inflammatory response in the body with *turmeric* and *bioflavonoids* (e.g. bitters, flax seeds, hawthorne, blueberries).

For burns from radiation, use *calendula* and *aloe* or *licorice plasters* (which have a steroid-like action).

Because radiation and its radioactive particles pose big cancer concerns, ongoing ingestion of *medicinal mushrooms* is the way to go. Just make sure both your mushrooms come from clean, radiation and chemical-free sources.

## Physical Support

With all of these different impacts and treatments, there are also physical things you can do to assist with the detoxing and organ support. Many folks recommend using the *sauna*. This can be fantastic, but you need to feel strong enough to spend time there. Also make sure you drink a lot of water when doing this.

## The Magic of Miso, Seaweed and Other Superfoods for Radiation Exposure

*Miso* is a traditional Japanese food produced by fermenting rice, barley or soybeans with salt and a special fungus called *kōjikin*. Some folks believe that miso can help treat radiation sickness, as people were fed miso after the Chernobyl nuclear disaster and the atomic bombings of Hiroshima and Nagasaki with positive results.[2] Japanese doctor Shinichiro Akizuki, who was the director of the Saint Francis Hospital from Nagasaki during World War II, conducted a study that indicated that miso helped protect against radiation sickness by binding and removing the radioactive elements from the body and decreasing the inflammation caused by radioactivity. In his experiment, rats were given both iodine-131 and cesium-134 (two elements commonly produced in nuclear reactor accidents), and those fed with miso had lower radiation levels in their blood. After the atomic bombing of Nagasaki, Dr. Akizuki fed his patients and hospital workers a diet of brown rice, miso, tamari soy soup, seaweeds, squash and sea salt and prohibited the consumption of sugar and sweets (as they suppress the immune system). The hospital he worked out was literally in the shadow of the blast, and supposedly none of Dr. Akizuki's patients suffered from radiation sickness.[3] In another study, the Japanese Cancer Institute found that folks who regularly consumed miso had a 50 percent lower incidence of cancer than those who did not eat miso.[4]

Eating a lot of clean, non-contaminated (so nowhere near the radiation plume or other contaminants) *seaweed* is a great safeguard before and after radiation exposure.[5] Kombu and bladderwrack, members of the kelp family, are some of the most radioprotective sea vegetables because they are high in a compound called fucoidan, a potent radioactive detoxifier, which is also known for inhibiting the growth of cancer cells.[6] Seaweeds like bladderwrack are also a good source of natural iodine (iodine-127), which the body prefers to radioactive iodine-131 ( a common radioisotope released in nuclear accidents and by nuclear power plants). It is most effective as a pre-exposure measure, as having sufficient natural iodine in your system can protect the thyroid gland from absorbing radioactive iodine. Also some brown algae seaweeds (like kelp, kombu, wakame  ☞

and bladderwrack) contain a compound called sodium alginate which has been known to reduce absorption of strontium-90 by 50 to 80 percent.[7] These seaweeds have also been found to be helpful in binding and eliminating other heavy metal contaminants (such as lead, cadmium and mercury) in the body.[8] The US Atomic Energy Commission recommended that adults consume two to three ounces (wet weight) of seaweed per week, or two tablespoons daily to protect from radiation toxicity, increasing that amount fourfold during or after direct exposure to radiation.[9]

*Spirulina,* a type of blue-green algae, has also been found to be quite helpful in dealing with the radiation. The results of studies conducted using spirulina in the treatment of children suffering from radiation sickness in Belarus following Chernobyl prompted the Belarus Ministry of Health to conclude that spirulina accelerates the evacuation of radionuclides from the human body.[10] A Russian report confirmed the Belurassian results a few years later, finding that "*Spirulina platensis* is an effective natural adsorbent of long-life radionuclides caesium-137, strontium-90, and potassium-40 and can lead to a reduction of radioactive contamination in children ages 10 to 16 years."[11] A chinese study also showed that an extract of spirulina had a protective effect from gamma radiation.[12] Spirulina and other algae contain high amounts of these metallo-thionine compounds which scientists think may strip the body of radioactive metals and protect against radiation damage.[13]

Foods such as barley/wheat grass and chlorella (a green freshwater microalgae you can get in pill form) contain chlorophyll, which can inhibit radiation and the metabolic activation of many carcinogens. Chlorella has also been touted to help the body detoxify heavy metal and other chemicals. There are many other foods and plants that help the body handle radiation and its impacts, such as tomatoes, beet juice, sea buckthorne, foods that are high in calcium and potassium (to assist in the excretion of radioactive particles, such as cesium-137) and foods that are high in beta-carotene, like squash, yams, carrots, spinach and chard.[14]

Bee pollen has also shown some promise in assisting folks in tolerating radiation. A study was conducted at the University of Vienna's Women's Clinic involving 25 women with inoperative uterine cancer receiving radiotherapy. The 15 who took 20 grams of bee pollen three times a day ☞

tolerated the radiation much better than the ten who took no bee pollen.[15] Finally, essential fatty acids (e,g, flax oil, fish oil, evening primrose oil) are believed to help with radiation exposure and in warding off cancer.[16]

Hot water baths with Epsom salts and essential oils are also great and lighter on the body. Daily movement and exercise are helpful. Lymphatic drainage massage, yoga, energy medicine and acupuncture are other healing modalities which work alongside the healing power of plants.

This information is just a tiny slice of what is out there in terms of the potential of the Earth to heal us as we take on the monumental task of earth repair. There are many plant and fungi allies out there which have much to offer, not only to the lands they seek to heal, but to we humans, who are as much a part of those landscapes as anything else is. Dandelions, milk thistle, yarrow, mullein, burdock, comfrey, red clover, plantain, fireweed, turkey tail and mushrooms of all sorts — these are some of nature's best healers, detoxers and disaster responders; they have adapted to live amongst the wreckage and bring it back to wholeness.

As part of your earth repair practice, consider growing and tending these powerful plant and mushroom healers in your backyard, in the community commons and in the wilds that they call home. Give thanks to them for their work, protect the places where they grow, help detoxify the environment around them, plant them everywhere and welcome them into your life!

# Conclusion:
# Final Words for a Fertile Way Forward

*If we will have the wisdom to survive,*
*to stand like slow growing trees on a ruined place, renewing,*
*enriching it ...*
*then a long time after we are dead*
*the lives our lives prepare will live*
*here ...*

— Wendell Berry, excerpt from
"Work Song, Part 2: A Vision"

THE WORK OF GRASSROOTS BIOREMEDIATION AND EARTH RE-
PAIR is a dynamic field that is simple, complicated, empowering
and overwhelming all at the same time, and it is one that demands
commitment, creativity and perseverance. If you take anything from
this book, take the passion to experiment and engage. Join forces with
the many bacterial, fungal, plant, animal and human allies to assist in
the healing of all those places that need love, care and justice.

I relish the many intersections, convergences and confluences you
find in the world of earth repair. Social justice flows into earth repair and
urban regeneration when abandoned trashed sites become community
berry farms, parks or food gardens that breathe life and vitality back
into our communities. Despair collides with hope, disaster clashes with
resilience and increasingly fertile resistance grows. The overwhelming
forces of climate chaos, nuclear accidents and catastrophic oil spills

meet the local, creative and courageous grassroots human and planetary healing responses. These are both the turbulent edges and the fierce centers where the work of earth repair is located, and where the juiciest, humblest and most inspiring responses are taking place as we unleash our true potential as healers, artists and visionaries.

Every time a grassroots bioremediation and earth repair project succeeds, it challenges the dominance of the industrial landscape. We can begin to change the stories of defeat and despair that make us accept the unacceptable and give up when we have yet to really begin trying. When weeds break through cracks in concrete, they remind us that living things are waiting for us to stop suffocating them. Weeds are patient, strong and will fight their way through with tenacity and resilience. Earth repair gives birth to spider webs of mycelia that rumble across the planet like little earthquakes, empowering more of us to wake up and throw down in irrepressible and brilliant ways.

Thank you to all the brave folks doing this work in all the many ways that it is being done, and in doing so making bold trails through the rubble and ruin back to those fertile places we can call home. This work is not easy, but it is absolutely necessary. In the wise words of Wendell Berry, "This is no paradisal dream. Its hardship is its reality."[1] May we continue to have the courage to tend to our dreams and re-wild our hearts for the sake of the survival of this planet. May we repair and regenerate that which has been wounded and destroyed, and may we leave these lands, waters and our communities more vibrant, healthy and alive for all those who follow in our footsteps.

# Appendix 1: Contaminants 101

WHEN IT COMES TO CLEANING UP CONTAMINATED SITES, it is important to understand what you are working with. Understanding what contaminants are present, how they act and their environmental and health impacts allows us proceed as effectively and safely as possible. Below are some of the more common heavy metal and organic chemical contaminants that you may find in the lands or waters you are working with. You can also find a list of websites with additional information in the endnotes.[1]

## Heavy Metals
### Arsenic (As)
Mining, smelting of nonferrous metals and the burning of fossil fuels are the major industrial processes that lead to arsenic contamination. Arsenic is also used in pesticides, herbicides, insecticides and as a preservative in pressure-treated lumber and wood products. In agriculture, organic arsenic compounds have been used in many agricultural insecticides and poisons, mainly on orchards and cotton fields. These compounds have also been used in livestock production as a controversial additive in animal food which supposedly stimulates growth and prevents disease. Other industrial products containing arsenic include lead-acid batteries, lead shots and bullets, some forms of glass, light-emitting diodes, paints, dyes, pyrotechnics, metals, pharmaceuticals, soaps and semiconductors.

Arsenic poisoning and long term exposure can lead to cancer of the skin, lungs, bladder, kidney, liver and prostate. Breathing high levels of inorganic arsenic can give you a sore throat or irritated lungs, and ingesting very high levels of arsenic can result in death. Exposure to lower levels in the short term can cause nausea and vomiting, swelling and redness of the skin, decreased production of red and white blood cells, abnormal heart rhythm, damage to blood vessels and a sensation of "pins and needles" in hands and feet.

### Cadmium (Cd)

Humans put cadmium into the environment by burning fossil fuels, coal and municipal solid waste, using phosphate fertilizers and producing iron and steel, cement and nonferrous metals. Cadmium is a minor component in most zinc ores and therefore is a byproduct of zinc production. It is also found in products such as batteries, paints and pigments, metal coatings and plastics. Cadmium can travel long distances before depositing in water or soil, and it binds strongly to soil particles.

Breathing high levels of cadmium can severely damage the lungs, and ingesting food or water with very high levels severely irritates the stomach, leading to vomiting and diarrhea. Long-term exposure to lower levels leads to a buildup of cadmium in the kidneys and possible kidney diseases, as well as lung damage and fragile bones.

### Lead (Pb)

Lead is currently used in building construction, lead-acid batteries, fusible alloys, pewters, bullets and shots, weights, as part of solders and as a radiation shield. In the construction industry, lead sheets are used as architectural metal in roofing material, cladding, flashing, gutters and gutter joints and on roof parapets. Lead is also frequently used in PVC (polyvinyl chloride) plastic, which coats electrical cords, as well as a coloring agent in some ceramic glazes. Some lead compounds are used as an additive for aviation fuel in piston-driven aircraft, and lead-based semiconductors are being used in photovoltaic cells (solar energy) and infrared detectors. Lead mining and smelting are also sources of environmental contamination globally.

Many older houses and buildings used lead-based paint, and as the paint ages, it chips off and mixes with the soil, which is why lead levels in residential soils tend to be highest closest to buildings. Lead was a component of the paint used on children's toys — now restricted in many countries. Prior to 1978, most fuels contained lead, and the exhaust from cars would deposit lead oxide onto soils near heavily trafficked roads. Most countries stopped using leaded gasoline as of 2007. Lead contamination may also come from leaded gasoline leaking from old underground storage tanks (at gas stations for example). Until the early 1970s, lead was also used for joining cast-iron water pipes and used as a material for small diameter water pipes. Any place where an older structure has burnt down will likely have lead contamination. Finally lead was used in pesticides before the 1950s, mainly for fruit orchards.

Lead damages the nervous system, especially in young children, as it is a neurotoxin that accumulates in both soft tissues and bones. Lead has been proven to permanently reduce the cognitive capacity of children at extremely low levels of exposure, causing learning and behavior disorders. It causes blood and brain disorders and is toxic to the heart, bones, intestines, kidneys and reproductive systems in both adults and children. Exposure can lead to small increases in blood pressure and can cause anemia. Exposure to high levels of lead can cause brain and kidney damage, as well as lead to miscarriages in pregnant women. Chronic, high-level exposures of lead have been found to reduce fertility in men. High levels of lead in the blood are associated with delayed puberty in girls.

### Chromium (Cr)

Natural sources of chromium compounds are present in the environment from the erosion of chromium-containing rocks and can be distributed by volcanic eruptions. The mined metal form of chromium is used for making steel, for chrome plating, dyes and pigments, leather tanning, brick making and wood preserving. Chromium is found in cement. Elevated levels of chromium can also be found in the air and soil near busy roads due to emissions from car brake lining and catalytic converters containing chromium. As a result of its widespread use, chromium compounds are often found in soil and groundwater at abandoned industrial sites. Primer paint containing chromium compounds are still widely used in aerospace and automobile refinishing businesses.

Health impacts from exposure to chromium include irritation to nose and skin, itching, nosebleed, sneezing, liver damage and cancer. Chromium is a known carcinogen.

## Mercury (Hg)

Mercury is used primarily for the manufacture of industrial chemicals or in electrical and electronic applications. Mercury contamination can occur through the production and the improper disposal (e.g. landfilling, incineration) of the waste stream of products that contain mercury such as batteries, fluorescent light bulbs, paint, thermometers, barometers, medical products, electrical switches, float valves, explosives, gold recovery, electrochemistry and tooth fillings. Coal plants are another major source of mercury contamination. Gold mining and production also accounts for mercury emissions and contamination, as well as nonferrous metal production and smelting. Oddly enough, mercury, as thimerosal, is widely used in the manufacture of mascara. Liquid mercury has been used as a coolant in some nuclear reactors, and before 1995 mercury was present in some herbicides and wood preservatives. It is also used in the production of chlorine and caustic soda at chlor-alkali process plants. Abandoned mercury mine processing sites often contain very hazardous waste piles of roasted cinnabar calcines, and water runoff from such sites is a recognized source of ecological damage. These mines and processing plants can be extremely toxic and contaminated places.

When mercury collects in rivers, lakes and streams and combines with rotting plants in oxygen-deficient conditions, it can turn into the more toxic methyl mercury. Even a very small amount of methyl mercury can poison all the fish in a pond or river. Methyl mercury in the environment is toxic for centuries. Conditions in the reservoirs of big dams can lead to formation of methyl mercury.

Mercury can cause skin rashes and impact the lungs and eyes. Methyl mercury is a potent neurotoxin capable of impacting neurological development in fetuses and young children and damaging the central nervous system of adults. Symptoms include impaired balance, numbness in hands and feet, muscle weakness, damage to hearing and speech and narrowing of the field of vision. In extreme cases, insanity, paralysis, coma and death can occur within weeks of the onset of symptoms. Exposure to mercury is most likely to occur at harmful quantities through consumption of fish contaminated with methyl mercury, as well as other food sources where mercury has bioaccumulated. Exposure to inorganic mercury also can happen from drinking contaminated water and touching contaminated water and soil.

Other heavy metals to watch out for and look up include: Aluminum (Al), Antimony (Sb), Barium (Ba), Beryllium (Be), Cobalt (Co), Copper (Cu), Iron (Fe), Lithium (Li), Manganese (Mn), Molybdenum (Mo), Nickel (Ni), Selenium (Se), Silver (Ag), Thallium (Tl), Tin (Sn), Uranium (U), Vanadium (V), Zinc (Zn).

## Organic Chemicals

### Persistent Organic Pollutants (POPs)

POPs are organic compounds that tend to persist in the environment for very long periods of time, where they biomagnify in the food chain and bioaccumulate in animals and humans. Because of this and their ability to travel through wind, air, soil and water, POPs are found all around the globe, even in places far from where they were produced. The main sources of POPs are the use of current or past pesticides and insecticides, as well as their use in industrial processes and in the production of pharmaceuticals, solvents, polyvinyl chloride and more.

As a result of POPs' toxicity, by the 1980s many restrictions and bans on the production and use of POPs came into effect. Concentrations of some POPs have decreased substantially; however, the contaminants already in the environment continue to circulate, and there are still many countries where these chemicals are being used as pesticides and in industrial processes. POPs are known to adversely affect the immune, reproductive, neurological, nervous, liver, skin and endocrine systems, as well as cause cancer. Even small exposures to POPs cause problems such as sterility and birth defects. Some POPs cause the body to become more chemically sensitive. Here are some common types of POPs.

### Polychlorinated Biphenyls (PCBs)

PCBs have been used as coolants and lubricants in transformers, capacitors and other electrical equipment. Products made before 1977 that may contain PCBs include old fluorescent lighting fixtures and electrical devices containing PCB capacitors — and old microscopes, hydraulic oils and caulking agents. These compounds were also used in industrial and commercial products such as rubber products, plastics, plasticizers in paints and hydraulic equipment. Even after banning, PCBs are released into the environment during their manufacturing process, from the improper disposal of products containing PCBs, from the incineration of some forms of waste or from leaks from electrical transformers. Because they are known to be very toxic, PCBs have been banned since the late 1970s in Canada and the USA and have been replaced in some cases by polybrominated diphenyl ethers (PBDEs). PBDEs also stay in our bodies for a long time, and can cause serious health problems such as damage to the brain and nerves.[2]

## Dioxins and Furans

Dioxins and furans are not intentionally manufactured, but form instead as byproducts of the manufacturing of other chlorinated chemicals, such as chlorinated wood preservatives and herbicides, and the chlorine bleaching of wood pulp for paper at pulp mills. These compounds may be generated by industrial processes or by combustion, including fuel burning in vehicles with leaded gasoline, municipal and medical waste incineration, metal smelting, open burning of trash, electrical transformer/capacitor fires and forest fires. Most dioxin is released when PVC plastic (commonly used to make pipes for water systems, as well as baby bottles, toys, food containers and other everyday products), bleached paper, coal, diesel fuel and other things that contain the chemical chlorine are burned. Dioxin is also released from metal smelting, cement making, papermaking and some pesticides. Dioxins and furans are not very mobile in the environment and tend to stick to soil and sediments. They can remain in soil for decades because they are incredibly slow to break down; they will be buried by accumulating sediments before they disappear.

## Dichlorodiphenyltrichloroethane (DDT) and other Pesticides

DDT is one of the most notorious and well-known pesticides/insecticides. It has been banned in the USA and Canada since the 1970s, but its use continues globally. It is present at many waste sites and releases from these sites could result in environmental contamination. Most DDT in the environment is the result of widespread past uses and is still present due to its slow breakdown. Other pesticides are now used, but they also have some pretty serious impacts on the environment and on human health. Pesticides contaminate the environment through leaks from production facilities and storage tanks, runoff from sprayed fields, agricultural spraying and disposal.

Other POPs to be aware of include the following pesticides: Aldrin, Alpha-hexachlorocyclohexane, Beta-hexachlorocyclohexane, Chlordane, Chlordecone, Dieldrin, Endosulfan, Endrin, Heptachlor, Hexachlorobenzene (HCB), Lindane, Mirex and Toxaphene. The following industrial chemicals and byproducts are also POPs: Hexabromobiphenyl (HBB), Hexabromocyclododecane (HBCD), Hexachlorobutadiene, Octabromodiphenyl ether, Pentabromodiphenyl ether (penta-BDE), Pentachlorobenzene, Perfluorooctane sulfonate (PFOS), Polychlorinated naphthalene (PCN) and short-chain chlorinated paraffins (SCCPs).

### Volatile Organic Compounds (VOCs)

According to the US Geological Survey, "many VOCs are human-made ... compounds of fuels, solvents, hydraulic fluids, paint thinners, and dry-cleaning agents commonly used in urban settings."[3] These compounds evaporate very quickly, hence the word "volatile" in their name. At elevated concentrations, VOCs affect the cardiovascular, neurological and nervous systems, and many are known carcinogens. The following VOCs, present in oil spills and other industrial processes, can pose a particular health risk to humans and the surrounding environment.

### Benzeneg

Benzene is a natural part of crude oil and gasoline, and is also used to make some types of rubbers, lubricants, plastics, dyes, detergents, drugs and pesticides. Industries that involve the use of benzene include the rubber industry, oil refineries, petroleum pipelines, coke and chemical plants, shoe manufacturers and gasoline-related industries and associated infrastructure. In most accidents and spills involving petroleum, benzene is definitely a concern. Natural sources of benzene include volcanoes and forest fires.

Benzene is a known carcinogen and can cause a rare form of kidney cancer and leukemia as well as other blood cancers. Short-term inhalation of high levels of benzene can be fatal, and low levels can cause drowsiness, dizziness, headaches, tremors, rapid heart rate, confusion or mental fog and unconsciousness. Eating foods or drinking water contaminated with high levels of benzene can cause vomiting, stomach irritation, dizziness, sleepiness, convulsions and death. Benzene damages the bone marrow and can lead to a decrease in red blood cells and anemia. It can also cause excessive bleeding and depress the immune system, resulting in a higher incidence of infections. It can impact the reproductive systems of men and women and cause birth defects such as spina bifida and anencephaly. When animals have been exposed to benzenes in studies, results have shown low birth weights, delayed bone formation and bone marrow damage.

## Ethyl Benzene

Ethyl benzene is found in coal tar and petroleum. It is used primarily to make the chemical styrene. It is used as a solvent, a constituent of asphalt and naphtha and is a constituent of synthetic rubber, fuels, paints, inks, carpet glues, varnishes, tobacco products and insecticides. It is a component of automotive and aviation fuels.

Acute exposure to ethyl benzene can cause eye, throat, nose, upper respiratory tract and mucous membrane irritation; chest constriction; redness and blistering of the skin. Neurological effects include dizziness, fatigue and lack of coordination. Animal studies have shown impacts to the central nervous system, pulmonary system and effects on the liver, kidneys and eyes. Chronic exposure to ethyl benzene can cause fatigue, headache, and eye and upper respiratory tract irritation, as well as drying, dermatitis and defatting of the skin.

## Toluene

Toluene occurs naturally in crude oil. It is also produced in the process of making coke from coal and gasoline and other fuels (such as jet fuel) from crude oil. Toluene is used in making paints, paint thinners, fingernail polish, lacquers, adhesives and rubber and in some printing and leather tanning processes. Low to moderate exposure to toluene can cause tiredness, confusion, weakness, drunken-type actions, memory loss, nausea, loss of appetite and loss of hearing and color vision. Toluene is also know to impact the cardiovascular system and the neurological/nervous system. Higher exposure levels can cause unconsciousness and death.

## Xylene

Xylene occurs naturally in petroleum and coal tar; it can catch on fire easily. It is found in small amounts in airplane fuel and gasoline. It is used in paints, paint thinners and varnishes. It is used also as a solvent and cleaning agent, and in the printing, rubber and leather industries.

Xylene exposure can damage the central nervous system, liver and other body systems. Signs and symptoms of acute exposure to xylene include headache, fatigue, irritability, lassitude, nausea, anorexia, flatulence, irritation of the eyes, nose and throat, issues with motor coordination and balance, flushing, redness of the face, a sensation of increased body heat, increased salivation, tremors, dizziness, confusion and cardiac irritability. Chronic exposure can cause central nervous system depression; conjunctivitis; dryness of nose, throat and skin; dermatitis; anemia; mucosal hemorrhage; bone marrow hyperplasia and kidney and liver damage.

## Trichloroethylene (TCE)

TCE is used primarily as a degreasing agent for metal and electronic parts; as an extractant for oils, waxes and fats; a solvent for cellulose esters and ethers; a dry-cleaning fluid (although it has largely been replaced since the 1950s by tetrachloroethylene); refrigerant and heat exchange fluid; fumigant; carrier agent in paints and adhesives; a scourant for textiles and as a feedstock for manufacturing organic chemicals. When first widely produced in the 1920s, its major use was to extract vegetable oils from plant materials such as soy, coconut and palm, as well as in coffee decaffeination. It has also been used in the medical field as an anesthetic. TCE can enter groundwater and surface water from industrial discharges or from improper disposal of industrial wastes at landfills. It can also be found in typewriter correction fluid, paint, spot removers, carpet-cleaning fluids, metal cleaners and varnishes.

When inhaled, TCE can cause central nervous system depression, liver and kidney damage. The symptoms of acute exposure can look similar to alcohol intoxication, beginning with a headache, dizziness and confusion and progressing with increasing exposure to unconsciousness. Respiratory and circulatory depression can eventually lead to death. TCE is believed to cause cancer (liver and kidney), leukemia, non-Hodgkin lymphoma as well as congenital heart defects.

There are many other VOCs (Tetrachloroethane, 1,2,4-Trichlorobenzene, Vinyl chloride) to be concerned about — those named above are just a few common ones.

### Polycyclic Aromatic Hydrocarbons (PAHs)

Polycyclic aromatic hydrocarbons are a group of over 100 different semi-volatile organic compounds that are formed during the incomplete burning of coal, oil and gas, garbage or other organic substances like tobacco or charbroiled meat. PAHs are found in coal tar, crude oil, creosote and roofing tar, but a few are used in medicines or to make dyes, plastics and pesticides. When coal is converted to natural gas, PAHs can be released, which is why some former coal-gasification sites may have elevated levels of PAHs. They are also found in incinerators, coke ovens and asphalt processing and use. They are also a major concern when it comes to human and environmental health impacts at oil spills, as they are present in crude oil. Although hundreds of PAHs exist, two of the more common ones are benzo(a)pyrene and naphthalene.

Polycyclic aromatic hydrocarbons can cause red blood cell damage that can lead to anemia; they can also suppress the immune system. Possible long-term health effects from exposure may include cataracts, kidney and liver damage and jaundice. Some polycyclic aromatic hydrocarbons are cancer-causing. Also, high prenatal exposure to PAHs is associated with lower IQ and childhood asthma, as well as low birth weight, premature delivery and heart malformations in babies.

## Radiation

Radioactive materials can poison the food, land and water for many generations and centuries. One of the main ways that radiation gets into our environment is from the nuclear industry, through its mining operations, power plants and waste disposal. Uranium, plutonium, strontium, cesium, tritium and iodine (among others elements) can be released via radioactively-contaminated water, air emissions, waste and accidents.

Many radioactive materials are produced by the military (e.g. depleted uranium shells). Radiation exposure is most common where weapons are made, tested and used, such as military bases and war zones. Radioactive metals are also used in some products such as electronics. In oil and gas drilling, radiation can be introduced into the environment by extraction processes that leave behind waste containing concentrations of naturally occurring radioactive material (NORM) from certain soils and rocks. Radioactive wastes from oil and gas drilling can be found in produced water, drilling mud, sludge, slimes or evaporation ponds and pits. It can also concentrate in the mineral scales that form in pipes (pipe scale), storage tanks or other extraction equipment. Radionuclides in these wastes are primarily radium-226, radium-228 and radon gas.

Radiation can cause cancer of the lungs, thyroid and blood, as well as diseases that affect the bones, muscles, nervous system, stomach and digestive system. Most exposure to harmful radiation occurs in small amounts over a long time, causing health problems to develop slowly. Nuclear accidents or explosions can cause death right away or within several weeks. People who survive six weeks after an explosion may recover for a while, but serious illness can return years later with many different forms of cancers and leukemia among other things. The effects of radiation can pass down from parents to children and grandchildren, as birth defects, cancers and other health problems. Early signs of radiation sickness include nausea, vomiting, diarrhea and fatigue. These signs may be followed by hair loss, burning feeling in the body, shortness of breath, swelling of the mouth and throat, worsening of tooth or gum disease, dry cough, heart pain, rapid heartbeat, permanent skin darkening, bleeding spots under the skin, anemia (pale or transparent skin, gums or fingernails) and death.

# Appendix 2:
# Conventional Remediation Techniques[1]

ERE ARE DESCRIPTIONS OF A FEW COMMON, CONVENTIONAL remediation technologies used by industry. Conventional remediation technologies can be *ex situ* or *in situ*. *Ex situ* methods involve excavating contaminated soils and then treating them at the surface, or removing them from the site for disposal. The transport of these hazardous loads of contaminated soil or water pose a risk to communities and the surrounding environment in the event of an accident or spill. *In situ* methods attempt to treat the contamination on-site and in place, without removing soil.

### Dig and Dump (Soil)

Contaminated soil is excavated and trucked off to a landfill. This common practice is also referred to as "Dig and Haul." There is also the "Dig and Burn" approach, where contaminated soil is taken to an incineration facility. When dealing with a contaminated river bottom or bay, dredging is often used.

### Pump and Treat (Groundwater)

Pumping and treating is one of the most common ways to deal with contaminated groundwater. A well is dug, and then groundwater is pumped up into a series of purification and treatment vessels using a submersible or vacuum pump. These vessels contain materials, such as activated carbon in the case of petroleum contamination, that are designed to adsorb contaminants. Chemical agents such as floccu-lants may also be used, followed then by sand filters to further filter and treat the groundwater. Once it is treated, the groundwater is then pumped back down.

## In Situ *Flushing (Soil and Groundwater)*

*In situ* flushing is used to help pump and treat groundwater and is one of the few methods that can help clean up non-aqueous phase liquids (NAPLs) on a site. NAPLs are chemicals (like solvents and heating oil) that exist as liquids but do not dissolve easily in water. *In situ* flushing is also known as "surfactant enhanced aquifer remediation" or "solubiliza-tion and recovery." Water along with a mixture of a surfactant/detergent or co-solvent is pumped down a well that is drilled in the contaminated area. The mixture helps dissolve NAPLs and also moves contamination to the next well where it is pumped up to be treated or recovered. It can take several months to several years for cleanup of a site. In terms of safety, there are some concerns around the use of surfactants and their impacts on the environment, as well as on clean-up workers.

## *Air Stripping (Groundwater)*

Air stripping is a chemical engineering technology used for the puri-fication of groundwater and wastewaters contaminated with volatile organic compounds (VOCs). Air stripping is easier and more effective at warmer temperatures.

## *Permeable Reactive Barriers (Groundwater)*

A permeable reactive barrier, also known as a "passive treatment wall," is used to clean up polluted groundwater. A long, narrow trench is dug in the path of the polluted groundwater and then filled with a reactive material (iron, limestone or carbon) to clean up the harmful contami-nants along with some sand to help with water flow. The filled trench or funnel is then covered with soil. The reactive material either traps contaminants to its surface, precipitates the contaminants out of the groundwater into the wall, changes the contaminants into a less harm-ful form and/or encourages microbes to eat the contaminants. The barriers can treat a range of different contaminants, such as organics, metals, inorganics and radionuclides. Permeable reactive barriers allow for on-site remediation, with no major excavation, pumping and disposal needed. It is a slow process, as groundwater flows slowly, and can take many years for cleanup to be complete.

## Membrane Filtration (Water)

This form of filtration separates contaminants from contaminated water by passing them through a semipermeable barrier or membrane that allows water and other low molecular weight chemicals to pass, while blocking contaminants with a higher molecular weight. These filtration processes can include microfiltration, ultrafiltration, nanofiltration and reverse osmosis.

## Soil Vapor Extraction and Air Sparging (Soil and Groundwater)

*Soil vapor extraction* and *air sparging* are often used to clean up both soil and groundwater simultaneously. This is also referred to as "Multi-Phase Extraction." Soil vapor extraction (SVE) removes chemicals, in the form of vapors, from contaminated soil above the water table, while air sparging focuses on removing vapors from polluted soil and groundwater below the water table. Both involve drilling wells into the contaminated zone and then using a vacuum to suck up the vapors. When air is pumped underground, the chemicals evaporate faster, which makes them easier to remove. Volatile organic compounds (VOCs), such as solvents and fuels, evaporate easily. The proper air pollution control equipment needs to be in place and monitored to make sure toxic vapors are not being released into the environment. The cleanup can take years, depending on the site and extent of contamination, though it can be speeded up by heating the soil by injecting steam or hot water into the wells.

## Mechanical Soil Aeration (Soil)

A simple method that agitates contaminated soil, using tilling or other means to volatilize contaminants that react with oxygen to evaporate.

## Soil Washing (Soil)

Soil washing "scrubs" soil to remove and separate the portion of the soil that is most polluted. Since chemicals tend to stick to more fine-grained soils like silt and clay, soil washing helps separate the silt and clay from the larger-grained, less-contaminated soil (sand and gravel).

Contaminated soil is dug up and sifted to remove rocks and debris. The sifted soil is placed in a machine called a scrubbing unit, where water and sometimes detergents are added. This mixture is passed through sieves, mixing blades and water sprays. Some of the contaminants may dissolve in the water or float. The polluted wash water is removed and cleaned up at a treatment plant. The separated silt and clay, which contain most of the pollution, is then tested for chemicals. If it needs further cleanup, it will be either washed again, cleaned using another remediation method or trucked to a landfill. The sand and gravel that settle to the bottom of the scrubbing unit are also tested for chemicals. If clean, they can be added back to the site. If pollution is still present, they will undergo more cleaning or remediation. Soil washing can take anywhere from several weeks to several months, depending on the site, the type of soil, the amount of pollution in the soil, the size of the scrubbing unit being used and the subsequent remediation strategies used to finish of the job.

## Solidifaction/Stabilization (Soil)

*Solidification* is a process that binds polluted soil or sludge and cements it into a solid block. *Stabilization* refers to changing the chemicals so they become less harmful or mobile. These two methods are often used together. A bonding agent is poured on contaminated soil to solidify and stabilize it in order to prevent contaminants from moving off-site via dust, leaching or runoff. The most commonly used bonding agents are cement, lime, limestone, fly ash, clay and gypsum.

Solidification/stabilization does not destroy the chemicals, and it doesn't help heal the site. It just covers and binds contamination. It is a widely used method used on contaminated brownfield sites and sites where hazardous waste is present.

This method can either be done *in situ* or *ex situ*, with most projects using the *ex situ* method of excavating the soil, machine mixing it with a bonding agent, then depositing it either at a specified location or back on the site. Soil can also be mixed in place, and then it is usually covered with a liner and clean soil or concrete. The process may take weeks or months to complete. Sites where this technique is used should be constantly monitored, as weathering (such as freezing, thawing, wetting and drying) and prolonged use can cause the bonding materials to erode, resulting in destabilization. It is also not an effective technique for organic contaminants or pesticides.

## In Situ *Chemical Oxidation (Soil and Groundwater)*

Chemical oxidation uses chemicals called oxidants to change pollution (like fuels, solvents and pesticides) in soil and groundwater into less harmful substances (like water and $CO_2$). Wells are drilled at different depths in the contaminated area, and oxidant is pumped into the ground where it mixes with the contaminants and begins the process of breaking it down. The oxidizing agents most commonly used for *in situ* oxidation are hydrogen peroxide, catalyzed hydrogen peroxide, potassium permanganate, sodium permanganate, sodium persulfate and ozone. While applicable to soil contamination, *in situ* oxidation is primarily used to remediate groundwater. *In situ* chemical oxidation has a fairly speedy turnaround time, with cleanups being measured in months instead of years. Some safety concerns with this method are that oxidants can corrode materials and burn skin, as well as explode if used under the wrong conditions.

## *Chemical/Solvent Extraction (Soil)*

Chemical/solvent extraction uses solvents to extract or remove contaminants that do not dissolve in water and stick to the soil (like PCBs, oil and grease). First, the soil is dug from the contaminated area and sifted to remove large debris. It is then placed in a machine called an extractor where it is mixed with a solvent. Sometimes soil has to go through this process several times before the contamination is remediated and soil can be returned to the site. Sometimes soil has to be heated to evaporate any remaining solvent. The bulk of the liquid solvent then travels to a machine called a separator where the contaminants are separated out of it and eventually the solvents are either destroyed or disposed of in a landfill. Extraction is a process that can take less than a year, but there are concerns about the chemicals used as solvents as they can have harmful impacts on the environment.

## *Metals Precipitation (Water and Groundwater)*

Precipitation of metals has long been the primary method of treating metal-laden industrial wastewater. By adjusting the pH, adding a chemical precipitant and flocculating the material, this process transforms dissolved contaminants in liquid form into an insoluble solid. The solids are then removed via sedimentation or filtration and disposed of.

For conventional clean-up methods for oil spills please see Chapter 8.

## Thermal Treatments

### Incineration (Soil)

Incineration is a heat-based technology that has been used for many years to burn and destroy contaminated materials at very high temperatures (1,600–2,200°F). Incineration is used to remediate soils contaminated with explosives and hazardous wastes, particularly chlorinated hydrocarbons, PCBs and dioxins. Incineration converts the waste into ash, flue gas and heat. The flue gases must be cleaned of pollutants before they are released into the atmosphere.

Incineration can decrease the volume of contaminated solid material by 90 percent, so that smaller amounts end up in the landfill. However, in destroying the chemicals it destroys the soil along with it (true of most of the thermal treatment methods). Incineration also necessitates, like landfilling, the transport of large amounts of contaminated materials to incinerators, putting communities and the environment at risk (unless it is in situ incineration). Incineration can sometimes produce more toxic and volatile chemicals than the original material, whose toxic vapors need to be trapped and treated before being disposed of. Incineration can also emit heavy metals. The ash produced through the incineration process is also highly toxic and needs proper disposal. A common concern is the significant amounts of dioxin and furan emissions, both considered to be serious health and environmental hazards. Incineration is a technology that is quick, but resource- and energy-intensive.

### Ex Situ and In Situ Thermal Desorption (Soil)

Thermal desorption is a process that removes harmful chemicals from soil (as well as sludge and sediment) by using heat to transform the chemicals into gases. The gases are collected by special equipment, changed back into liquids and/or solid materials and disposed of. The clean soil can then be returned to the site, or if it still contains contaminants, it will be landfilled or put through another cleaning process. Thermal desorption differs from incineration as it seeks to change the contaminants into gases for separation and capture, instead of destroying the contaminants. Considered a relatively quick process, it can take up to several weeks on small sites, and up to several years for larger contaminated areas.

## Vitrification (Soil)

Vitrification is a process that traps harmful chemicals in a solid block of glass-like material. Electric power heats up the soil enough to melt it. That's right, melts the soil at elevated temperatures (2,900 to 3,650°F). Four rods, called electrodes, are drilled into the contaminated area and a current is passed between them. The high temperature attained in the process destroys or removes organic materials. Radionuclides and most heavy metals are retained within the now solidified, melted soil block. When the melted soil cools, the vitrification product is supposedly a chemically stable, leach-resistant, glass and crystalline material similar to obsidian or basalt rock. Its volume shrinks causing the area to sink slightly, and then it turns into a solid block of glass-like material, electrodes included. The heat destroys some of the chemicals; others evaporate and are captured by a giant hood placed about the area during vitrification. The glass-like block of melted soil is left in place on-site, where it is monitored to make sure contaminants are not being released. Vitrification is one of the quicker clean-up methods, taking from several weeks to several months.

Other thermal treatments used include Hot Gas Decontamination, Plasma High Temperature Recovery and Pyrolysis. Thermal treatments offer quick clean-up times but are typically the most costly and machine-dependent technologies.

## Big Bioremediation

### Bioaugmentation and Biostimulation (Soil and Groundwater)

At its core, bioremediation works with natural processes, allowing for polluted soil and groundwater to be cleaned on-site without having to dispose of it in a landfill. In conventional remediation, bioremediation is thought of as "green remediation," but it can look quite different than grassroots bioremediation. *Biostimulation* depends on indigenous microorganisms to break down contaminants, and therefore seeks to stimulate their growth and to create favorable conditions for them to live. Generally, this means providing some combination of oxygen, nutrients (in the form of fertilizers), soil amendments and moisture, controlling temperature and pH. Sometimes compost is brought in, other times biosolids are used (they seem to be gaining popularity). If the conditions are not correct, the microorganisms die, grow slowly or create more harmful byproducts. *Bioaugmentation* refers to a process in which microorganisms that have been adapted to degrade specific contaminants are applied to enhance the bioremediation process.

Depending on the site and location of the contamination, it may be difficult to create the right conditions for effective bioremediation, for example in cold weather or in dense or waterlogged soil. In cases like these, the soil is dug up and then fertilizer and oxygen is added to it. This is either done outside or it is done within a special tank. VOCs and fuels are remediated effectively in this way, pesticides to a lesser extent. Bioremediation is also used for groundwater contamination. These technologies may take anywhere from several months to several years for cleanup. There can be serious environmental impacts and human health impacts from using certain fertilizers. If done improperly, fertilizer application can throw the microecology of a site out of whack. There is also some work being done in this field that involves breeding and working with bacteria in labs and then importing them onto a site, which is not always ideal and raises some concerns.

### Monitored Natural Attenuation (Soil and Groundwater)

Monitored natural attenuation (MNA) is basically "letting nature handle it." It depends on microorganisms to break contaminants down, chemicals to evaporate, that the soil will bind some contaminants and that some will become diluted in groundwater. Handling a site via natural attenuation is often done after the main toxic load has been removed, and there are only small amounts left. The sites must be constantly monitored; cleanup by natural forces can take several years to several decades or more, depending on the contaminants and the site.

### Land Farming (Soil and Sludge)

Considered a bioremediation technique, land farming may sound like a green and innocuous way to go, but names can be deceiving. For affected communities, it is quite a controversial remediation method whereby contaminated soils, sediments or sludge are dumped at a different, uncontaminated site and incorporated into the soil surface. The soil is turned over/tilled periodically to aerate the mixture to speed breakdown of contaminants, evaporate volatile gases and stimulate aerobic microbial activity within the soils through aeration and/or the addition of minerals, nutrients and moisture. Land farming has been used to break down petroleum hydrocarbons and other less volatile, biodegradable contaminants. The more chlorinated or nitrated the compound, the more difficult it is to degrade. Land farming is not effective at breaking down metals, but they can become immobilized through this process. Contaminants that have been successfully treated include diesel fuel, polycyclic aromatic hydrocarbons, oily sludge, wood-preserving wastes and certain pesticides.

Why the controversy? Because this method involves taking contaminated soils and mixing it into non-contaminated soil, which can actually spread out the contamination and result in new sacrifice zones with polluted soil, air, surface and groundwater, via vapors, dust, off-gassing, runoff and leaching. Land farming must be properly managed to avoid expanding contamination. By mixing contaminated soil with uncontaminated soil, this method requires a large area; it also initially increases the volume of contaminated material.

# Appendix 3: Metric Conversion Table

| If you know | Multiply by | To Find |
|---|---|---|
| inches | 25.4 | millimeters |
| inches | 2.54 | centimeters |
| feet | 0.3048 | meters |
| yards | 0.9144 | meters |
| miles | 1.609 | kilometers |
| fluid ounces | 28.4 | milliliters |
| gallons | 4.546 | liters |
| ounces | 28.350 | grams |
| pounds | 0.4536 | kilograms |
| short tons | 0.9072 | metric tons |
| acres | 0.4057 | hectares |
| square inches | 6.452 | square centimeters |
| square yards | 0.836 | square meters |
| cubic yards | 0.765 | cubic meters |
| square miles | 2.590 | square kilometers |

Temperature

To convert degrees Fahrenheit to degrees Celsius: $(°F - 32) \div 1.8$

# Endnotes

## Chapter 2: Earth Repair and Grassroots Bioremediation

1. Scott Kellogg and Stacy Pettigrew. *Toolbox for Sustainable City Living: A Do-It-Ourselves Guide.* South End, 2008.
2. Robert Rawson, personal interview, Graton, CA, June 15, 2012.
3. Mark Lakeman, Skype interview, March 27, 2012.

## Chapter 3: Getting Started

1. Starhawk, personal interview, Cazadero, CA, June 13, 2012.
2. Rawson, personal interview.

## Chapter 4: Microbial Remediation

1. Ja Schindler, excerpt from "Focusing on Soil Remediation and 'Pre-Mediation' with Fungi," see Chapter 6.
2. Marika Smith, personal interview, February 28, 2012.
3. R.E. Pettit. "Organic Matter, Humus, Humate, Humic Acid, Fulvic Acid and Humin: Their Importance in Soil Fertility and Plant Health." Texas A&M University, February 2008. [online]. [cited December 21, 2012]. greenhumus.com.cn/photo/humic%20acid%20details.pdf.
4. Elaine Ingham, Skype interview, February 15, 2012.
5. Compiled using Kellogg and Pettigrew. *Toolbox for Sustainable City Living* and Joseph Jenkins. *The Humanure Handbook*, 2nd ed., 1999. [online]. [cited May 4, 2012]. weblife.org/humanure/chapter3_7.html.
6. Scott Kellogg, Skype interview, May 4, 2012.
7. David Holmgren, Skype interview, May 11, 2012.
8. J. Tharakan, A. Addagada, D. Tomlinson and A. Shafagati. "Vermicomposting for the bioremediation of PCB congeners in SUPERFUND

site media." *Earth Worm Digest*, December 4, 2005. [online]. [cited May 8, 2012]. wormdigest.org/content/view/186/2/.

9. Betsy Lynch. "Vermiculture Rehab." Urban Farm, June 14, 2010. [online]. [cited May 8, 2012]. urbanfarmonline.com/urban-farm-news/2010/06/14/worms-reclaim-soil.aspx.

10. Jeff Lowenfels and Wayne Lewis. *Teaming with Microbes: The Organic Gardener's Guide to the Soil Food Web*, rev. ed. Timber, 2010, p. 98.

11. Ingham, Skype interview.

12. Ja Schindler, email interview, December 3, 2012.

13. Kellogg and Pettigrew, *Tools for Sustainable City Living*, p. 185.

14. Matthew Slaughter, Earthfort Compost Tea Webinar, December 20, 2012.

15. Rawson, personal interview.

16. Ibid.

17. Ibid.

18. Lowenfels and Lewis, *Teaming with Microbes*, p. 21.

19. Nance Klehm, Skype interview, October 30, 2012.

## Chapter 5: Phytoremediation

1. you are the city. *Brownfields to Greenfields: A Field Guide to Phytoremediation*. youarethecity.com, 2011. [cited March 13, 2013].

2. Ibid.

3. USDA Natural Resources Conservation Service. *Web Soil Survey*. [online]. [cited November 27, 2012]. websoilsurvey.nrcs.usda.gov/app/HomePage.htm.

4. Leon V. Kochian. "Phytoremediation: Using Plants To Clean Up Soils." USDA Agricultural Research Service website, August 13, 2004. [online]. [cited November 27, 2012]. ars.usda.gov/is/ar/archive/jun00/soil0600.htm.

5. Compiled using: G.S. Bañuelos et al. "Phytoextraction of selenium from soils irrigated with selenium-laden effluent." *Plant and Soil*, Vol 224#2 (2000), pp. 251–258; Rufus L. Chaney et al. "Phytoremediation of Soil Metals." *Current Opinions in Biotechnology* Vol. 8 (1997), pp. 279–284. [online]. [cited December 22, 2012]. soils.wisc.edu/~barak/temp/opin_fin.htm; Tesema Chekol et al. "Plant-Soil-Contaminant Specificity Affects Phytoremediation of Organic Contaminants." *International Journal of Phytoremediation*, Vol. 4#1 (2002), pp. 17–26. [online]. [cited December 22, 2012]. environmentalhealthclinic.net/wp-content/uploads/2007/11/plant-soil-contaminant-specificity-affects-phytoremediation-of-organic-contaminants.pdf; Joseph Devinny et al. *Phytoremediation with Native Plants*. The Zumberg Fund for Innovation, March 15, 2005. [online]. [cited April 16, 2012]. urbanwildlands.org/Resources/SpiralingRootsZumberge.pdf;

Steve Diver. "Phytoextraction of Arsenic with Fern, Heavy Metal Reuse: Online Interview with Rufus Chaney." *Sustainable Agriculture and Research Network*. March 2, 2001. [online]. [cited December 21, 2012]. lists.ifas. ufl.edu/cgi-bin/wa.exe?A2=ind0103&L=sanet-mg&P=R2570&I=-3; Slavik Dushenkov et al. "Phytoremediation of Radiocesium-Contaminated Soil in the Vicinity of Chernobyl, Ukraine." *Environmental Science and Technology* 1999. Vol 33#3 (1999), pp. 469–475. [online]. [cited May 7, 2012]. lib3.dss.go.th/fulltext/Journal/Environ%20Sci.%20Technology 1998-2001/1999/no.3/3,1999%20vol.33,no3,p469-475.pdf; Srujana Kathi and Anisa B. Khan. "Phytoremediation approaches to PAH contaminated soil." *Indian Journal of Science and Technology*, Vol. 4#1 (January 2011), pp. 56–63. [online]. [cited December 22, 2012]. indjst.org/ archive/vol.4.issue.1/Jan11anisakhan-14.pdf; Stephen C. McCutcheon and Jerald L. Schnoor, eds. *Phytoremediation: Transformation and Control of Contaminants*. Wiley-Interscience, 2003; D. Parker et al. "Selenium phytoremediation potential of *Stanleya pinnata*." *Plant and Soil*, Vol. 249#1 (February 2003), pp.157–165; youarethecity, *Brownfields to Greenfields*.

6. Ken Roseboro. "Poplar Trees Root Out Pollution." *American News Service*, September 28, 2000. [online]. [cited October 14, 2012]. berkshire publishing.com/ans/HTMView.asp?parItem=S031000645A.

7. Nance Klehm, email interview, December 20, 2012.

8. Check out Kaja's blog at newyork.thecityatlas.org/category/atlas-lab/ brownfield-remediation/. [cited march 13, 2013].

9. Jay Rosenberg, phone interview, December 20, 2012.

10. The US Environmental Protection Agency. *The National LUST Clean Up Backlog: A Study in Opportunities*. [online]. [cited December 8, 2012]. www.epa.gov/oust/cat/backlog.html.

11. Sadra Heidary-Monfared. "Table 2-3 Maximum acceptable limits (ppm) of heavy metals in soil in various countries and regions for residential and recreational land uses." *Community Garden Heavy Metal Study*. Ecology Action Centre, Environment Canada, Nova Scotia Agricultural College and Nova Scotia Environmental Network, January 2011, p. 15. [online]. [cited December 8, 2012]. ecologyaction.ca/files/images/file/Community%20 Garden%20Heavy%20Metal%20Contamination%20Study.pdf.

12. Arthur Craigmill and Ali Harivandi. *Home Gardens and Lead: What You Should Know about Growing Plants in Lead Contaminated Soil*. University of California Agriculture and Natural Resources, Publication 8424. September 2010. [online]. [cited March 14, 2012]. anrcatalog.ucdavis. edu/pdf/8424.pdf.

13. Meg Perry Healthy Soil Project. *The New Orleans Residents' Guide To Do It Yourself Soil Clean Using Natural Processes*. Common Ground Relief, 2006.

14. US Environmental Protection Agency. "The Fishbone Project for a Lead Safe Community." *West Oakland Lead Cleanup Update,* January 2012. [online]. [cited December 23, 2012]. epaosc.org/sites/5604/files/Fishbone1_12%20373kb.pdf.

15. D.R. Ownby et al. "Lead and zinc bioavailability to *Eisenia fetida* after phosphorous amendment to repository soils." *Environmental Pollution,* Vol. 136#2 (July 2005), pp. 315–21. [online]. [cited December 21, 2012]. http://www.ncbi.nlm.nih.gov/pubmed/15840539.

16. David E. Stilwell and John F. Ranciato. "Use of Phosphates to Immobilize Lead in Community Garden Soils." The Connecticut Agricultural Experiment Station, Bulletin 1018, 2008. [online]. [cited December 21, 2012]. ct.gov/caes/lib/caes/documents/publications/bulletins/b1018.pdf.

17. Andrew Butcher, phone interview, April 24, 2012.

## Chapter 6: Mycoremediation

1. Brian Spooner and Peter Roberts. *Fungi.* Collin, 2005.

2. G.M. Gadd, ed. *Fungi in Bioremediation (British Mycological Society Symposia).* Cambridge, 2008.

3. Harbhajan Singh. *Mycoremediation: Fungal Bioremediation.* Wiley-Interscience, 2006.

4. Paul Stamets. *Mycelium Running: How Mushrooms Can Help Save the World.* Ten Speed, 2005.

5. Amazon Mycorenewal Project website. [online]. [cited November 7, 2012]. amazonmycorenewal.org.

6. Radical Mycology website. [online]. [cited November 7, 2012]. radicalmycology.wordpress.com.

7. US Department of Health and Human Services. Agency for Toxic Substances and Disease Registry. [online]. [cited November 19, 2012]. atsdr.cdc.gov.

8. Rawson, personal interview.

9. Mia Rose Maltz, email interview, June 22, 2012.

10. Paul Stamets, Skype interview, February 23, 2012.

11. Paul Stamets. *Growing Gourmet and Medicinal Mushrooms,* 3rd ed. Ten Speed, 2000; Paul Stamets and J.S. Chilton. *The Mushroom Cultivator: A Practical Guide to Growing Mushrooms at Home.* Agarikon, 1983.

12. Scott Koch, Skype interview, October 31, 2012.

## Chapter 7: The Art of Healing Water

2. Craig Sponholtz, telephone interview, May 30, 2012.

3. Craig Sponholtz. *10 Guiding Principles for Watershed Restoration Regenerative Water Harvesting,* October 8, 2012. [online]. [cited December

3, 2012]. floodwateragroecology.com/blog/uncategorized/10-guiding-principles-for-watershed-restoration-regenerative-water-harvesting/.

4.  Sponholtz, phone interview.

5.  Adrian Buckley, personal interview, May 3, 2012.

6.  Brad Lancaster, telephone interview, May 29, 2012.

7.  Phil Nauta. "Rain Harvesting Into Soil — Putting Your Rainwater Barrel To Shame." Smiling Gardener website, October 25, 2010. [online]. [cited June 29, 2012]. smilinggardener.com/water-management/rain-harvesting.

8.  Lancaster, phone interview.

9.  The City of Calgary. "Shepard stormwater diversion project." The City of Calgary website, 2012. [online]. [cited June 29, 2012]. calgary.ca/UEP/Water/Pages/construction-projects/Stormwater-quality-retrofit-program/Shepard-stormwater-diversion-project/Shepard-Stormwater-Diversion-Project.aspx.

10. Ducks Unlimited. "Wetlands and Grassland Habitat." [online]. [cited January 3, 2013]. ducks.org/conservation/habitat/page2.

11. T.M. Samuels. "Aquatic Plants That Purify Water." Livestrong.com website, August 4, 2010. [online]. [cited June 29, 2012]. livestrong.com/article/193311-aquatic-plants-that-purify-water/#ixzz1uX1qGaI5.

12. Ibid.

13. Ibid.

14. M.D. Ansal, A. Dhawan and V.I. Kaur. "Duckweed based bio-remediation of village ponds: An ecologically and economically viable integrated approach for rural development through aquaculture." *Livestock Research for Rural Development*, Vol. 22#7 (July 1, 2010). [online]. [cited June 29, 2012]. lrrd.org/lrrd22/7/ansa22129.htm; Clemson University Cooperative Extension. "Duckweed." Clemson University, 2012. [online]. [cited June 29, 2012]. clemson.edu/extension/horticulture/nursery/remediation_technology/constructed_wetlands/plant_material/duckweed.html; Wikipedia. "Lemnoideae." Wikipedia, June 28, 2012. [online]. [cited June 29, 2012]. en.wikipedia.org/wiki/Lemnoideae.

15. Samuels, "Aquatic Plants That Purify Water."

16. Shao-Wei Liao and Wen-Lian Chang. "Heavy Metal Phytoremediation by Water Hyacinth at Constructed Wetlands in Taiwan." *Journal of Aquatic Plant Management*, Vol. 42 (2004), pp. 60–68. [online]. [cited June 29, 2012]. apms.org/japm/vol42/v42p60.pdf; Rouf Ahmad Shah et al. "Water Hyacinth (*Eichhornia Crassipes*) as a Remediation Tool for Dye-Effluent Pollution." *International Journal of Science and Nature*, Vol. 1#2 (2010), pp. 172–178. [online]. [cited June 29, 2012]. scienceandnature.org/IJSN_V1(2)_D2010/IJSN_V1(2)_15.pdf; Clemson University Cooperative

Extension. "Water Hyacinth." Clemson University, 2012. [online]. [cited June 29, 2012]. clemson.edu/extension/horticulture/nursery/remediation_technology/constructed_wetlands/plant_material/water_hyacinth.html.

17. Samuels, "Aquatic Plants That Purify Water."

18. Sandy Coyman and Keota Silaphone. "Rain Gardens in Maryland's Coastal Plain." Worcester County Commissioners and the Maryland Department of the Environment, p. 2. [online]. [cited January 3, 2013]. aacounty.org/DPW/Highways/Resources/Raingarden/Rain_Gardens_MD_Coastal_Plain.pdf .

19. Cleo Woelfle-Erskine and Apryl Uncapher. *Creating Rain Gardens: Capturing the Rain for Your Own Water-Efficient Garden.* Timber, 2012; CloudCatching website. [online]. [cited December 4, 2012]. Cloud Catching.org.

20. Melbourne Water. "Healthy Waterways Raingardens." [online]. [cited June 29 , 2012]. raingardens.melbournewater.com.au/.

21. Stamets, *Mycelium Running,* p. 58.

22. Kellogg and Pettigrew, *Toolbox for Sustainable City Living;* Scott Kellogg and Stacy Pettigrew, illustrations by Juan Martinez. "How to build a floating trash island." *Low-tech Magazine.* [online]. [cited June 29, 2012]. lowtechmagazine.com/how-to-build-a-floating-trash-island.html.

23. Wikipedia. "Daylighting (streams)." Wikipedia. [online]. [cited June 29, 2012]. en.wikipedia.org/wiki/Daylighting_(streams).

24. Berklee Lowrey-Evans. "Letting the Xingu Run Freely." *International Rivers,* June 15, 2012. [online]. [cited June 29, 2012]. internationalrivers. org/blogs/244/letting-the-xingu-run-freely.

## Chapter 8: Oil Spills I: The Anatomy of an Environmental Disaster

1. Paul Horsman, Skype interview, February, 13, 2012.

2. Nir Barnea. "Health and Safety Aspects of In-situ Burning of Oil." US National Ocean and Atmospheric Association. [online]. [cited December 30, 2012]. response.restoration.noaa.gov/sites/default/files/health-safety-ISB.pdf.

3. "Scientific Support Provided by Fisheries and Oceans Canada During the Gulf of Mexico Oil Spill Response Operations Boosts Canada's Oil Spill Preparedness." Department of Fisheries and Oceans Canada website, April 4, 2012. [online]. [cited December 30, 2012]. dfo-mpo.gc.ca/science/publications/article/2012/04-04-12-eng.html.

4. Horsman, Skype interview.

5. Anita M. Burke, Skype interview, February 9, 2012.

6. Horsman, Skype interview.

7. "Oilsands tailings ponds kill more ducks." CBC News, October 26, 2010.

[online]. [cited December 30, 2012]. cbc.ca/news/canada/edmonton/story/201 0/10/26/edmonton-more-ducks-tailings-pond.html.

8. Burke, Skype interview.

9. Dahr Jamail. "Gulf spill sickness wrecking lives." Al Jazeera, March 9, 2011. [online]. [cited December 30, 2012]. aljazeera.com/indepth/features/2011/03/201138152955897442.html; Anne Casselman. "A Year After the Spill, 'Unusual' Rise in Health Problems." *National Geographic News*, April 20, 2011. [online]. [cited December 30, 2012]. news.nationalgeographic.com/news/2011/04/110420-gulf-oil-spill-anniversary-health-mental-science-nation/; Joanna Zelman. "Gulf Oil Spill Cleanup Workers Report Mysterious Illnesses Year After Disaster." *The Huffington Post*, updated June 18, 2011. [online]. [cited December 30, 2012]. huffingtonpost.com/2011/04/18/gulf-oil-spill-health-cleanup-workers_n_850486.html.

10. "BP Tells Fishermen Working On The Oil Spill That They Will Be Fired For Wearing A Respirator." Louisiana Environmental Action Network, June 1, 2010. [online]. [cited December 30, 2012].leanweb.org/our-work/water/bp-oil-spill/bp-tells-fishermen-working-on-the-oil-spill-that-they-will-be-fired-for-wearing-a-respirator.

11. Jamail. "Gulf spill sickness wrecking lives;" Michael R. Harbut and Kathleen Burns. "Gulf Oil Spill Health Hazards." *Sciencecorps*, June 14, 2010. [online]. [cited December 30, 2012]. sciencecorps.org/crudeoil hazards.htm.

## Chapter 9: Oil Spills II: Tools for the Grassroots Bioremediator

1. Lisa Craig Gauthier, email interview, July 8, 2012.

2. Paul Stamets. "The Petroleum Problem," June 3, 2010. [online]. [cited December 10, 2012]. fungi.com/blog/items/the-petroleum-problem.html.

3. Rawson, personal interview.

4. Maltz, email interview.

5. S. Thomas et al. "Mycoremediation of Aged Petroleum Hydrocarbon Contaminants in Soil." Research Report WA-RD 464.1, Pacific Northwest National Laboratory for the Washington State Department of Transportation, November 1998. [online]. [cited December 15 2012].wsdot.wa.gov/research/reports/fullreports/464.1.pdf.

6. Burke, Skype interview.

7. Maltz, email interview.

8. Gauthier, email interview.

9. Maltz, email interview.

10. Stamets, Skype interview.

11. International Bird Rescue. "How Oil Affects Birds" [online]. [cited March

17, 2012]. bird-rescue.org/our-work/research-and-education/how-oil-affects-birds.aspx.

12. Horsman, Skype interview.

13. International Bird Rescue. "Our Process for Helping Oiled Birds" [online]. [cited March 17, 2012]. bird-rescue.org/our-work/aquatic-bird-rehabilitation/our-process-for-helping-oiled-birds.aspx.

14. Wildpro.com. "Washing, Rinsing and Drying Mammals." Cleaning Oiled Wildlife. [online]. [cited January 4, 2013]. wildpro.twycrosszoo.org/S/00Man/Wlife_OilspillTech/11oilspill_washing.htm#mammals.

15. Oiled Wildlife Care Network. "Rescue and Treatment for Oiled Animals." UC Davis Veterinary Medicine. [online]. [cited December 15, 2012]. vetmed.ucdavis.edu/owcn/oiled_wildlife/rescue_and_treatment.cfm.

16. S. Shafir, J. Van Rijn and B. Rinkevich. "Short and long term toxicity of crude oil and oil dispersants to two representative coral species." Israel Oceanographic and Limnological Research, National Institute of Oceanography. *Environmental Science and Technology*, Vol. 41#15 (2007), pp. 5571–4.; R. Rico Martinez et al. "Synergistic toxicity of Macondo crude oil and dispersant Corexit 9500A® to the *Brachionus plicatilis* species complex (Rotifera)." Universidad Autónoma de Aguascalientes, Mexico. *Environmental Pollution*, Vol 173 (2013), pp. 5–10.

17. US Environmental Protection Agency. "Oil Spill: How Is Air Quality Affected?" August 19, 2010. [online]. [cited May 9, 2012]. epa.gov/enbridgespill/pdfs/enbridge_fs_20100819aq.pdf; US Fish and Wildlife Service (Midwest Region). "Michigan Oil Spill Response: Frequently Asked Questions." August 10, 2010. [online]. [cited May 9, 2012]. fws.gov/midwest/oilspill/FAQs.html.

18. DuPont. *Tychem® TK*. [online]. [cited May 8, 2012]. www2.dupont.com/personal-protection/en-us/dpt/tychem-tk.html.

19. Check the online General Personal Protective Equipment Sampling to get an idea of what gear the US government suggests for clean-up workers. As government agencies are subject to industry lobbying, certain things, in my option, are downplayed, but this is definitely helpful to see what standards are out there and what is recommended for different oil spill clean-up activities: US Department of Labor. Occupational Health and Safety Administration. *General Personal Protective Equipment Sampling Matrix*. [online]. [cited December 11, 2012]. osha.gov/oilspills/oil_ppematrix.html.

20. Stephen Hume. "Pipeline spills are not the exception in Alberta, they are an oily reality." *Vancouver Sun*, June 13, 2012. [online]. [cited August 4, 2012]. vancouversun.com/news/Hume+Pipeline+spills+exception+Alberta+they+oily+reality/6778601/story.html#ixzz2FS8D5Oui.

21. Richard Girard and Tanya Roberts Davis. *Out on the Tar Sands Mainline: Mapping Enbridges' Web of Pipelines.* Polaris Institute, 2010. [online]. [cited January 8, 2013]. tarsandswatch.org/files/Updated%20Enbridge%20Profile.pdf.

22. Anthony Swift et al. "Pipeline and Tanker Trouble: The Impact To British Columbia's Communities, Rivers, and Pacific Coastline from Tar Sands Oil Transport." Natural Resources Defense Council, The Pembina Institute and The Living Oceans Society, November 2011. [online]. [cited December 11, 2012]. nrdc.org/international/files/PipelineandTankerFS.pdf.

23. M. Pengelly et al. *West Coast Marine Weather Hazards Manual.* Environment Canada, 1992, p. 113.

24. Martha Stanbury et al. "Acute Health Effects of the Enbridge Oil Spill." Michigan Department of Community Health, November 2010. [online]. [cited March 10, 2012]. michigan.gov/documents/mdch/enbridge_oil_spill_epi_report_with_cover_11_22_10_339101_7.pdf.

25. Mitchell Anderson. "Spill from Hell: Diluted Bitumen." *The Tyee*, March 5, 2012. [online]. [cited May 9, 2012]. thetyee.ca/News/2012/03/05/Diluted-Bitumen/.

26. Swift, "Pipeline and Tanker Trouble."

## Chapter 10: Nuclear Energy and Remediating Radiation

1. Winifred Bird. "Fukushima nuclear cleanup could create its own environmental disaster." *The Guardian*, January 9, 2012. [online]. [cited April 8, 2012]. guardian.co.uk/environment/2012/jan/09/fukushima-cleanup-environmental-disaster.

2. S. Manley et al. "Canopy-Forming Kelps as California's Coastal Dosimeter: $^{131}$I from Damaged Japanese Reactor Measured in *Macrocystis pyrifera*" Department of Biological Sciences, California State University, Long Beach, March 6, 2012. *Environmental Science and Technology*, Vol.46#7 (2012), pp 3731–3736; Alex Roslin. "Fukushima fallout hit Canada: Radioactive iodine detected in Canadian rainwater but Harper gov't kept public in in the dark." *Montreal Gazette*, January 17, 2012. [online]. [cited April 8, 2012]. sierraclub.ca/en/water-campaign/in-the-news/fukushima-fallout-hit-canada.

3. The McGraw-Hill Companies. "Phytoremediation: Using Plants to Clean Soil." Botany Global Issues Map, 2000. [online]. [cited April 10, 2012]. mhhe.com/biosci/pae/botany/botany_map/articles/article_10.html.

4. H. Vandenhove and M. Van Hees. "Fibre crops as alternative land use for radioactively contaminated arable land." *Journal of Environmental Radioactivity*, Vol. 81 (2005), pp.131–141; McGraw-Hill, "Phytoremediation: Using Plants to Clean Soil."

5. R.J. Fellows et al. "100-N Area Strontium-90 Treatability Demonstration Project: Phytoextraction Along the 100-N Columbia River Riparian Zone — Field Treatability Study." Document PNNL-19120 prepared for US Department of Energy by Pacific Northwest National Laboratory, January 2010. [online]. [cited April 11, 2012]. pnl.gov/main/publications/external/technical_reports/PNNL-19120.pdf; K.M. Thompson et al. "An Innovative Approach for Constructing an In-Situ Barrier for Strontium-90 at the Hanford Site, Washington — 9325." WM2009 Conference, March 1–5, 2009. [online]. [cited April 10, 2012]. wmsym.org/archives/2009/pdfs/9325.pdf.

6. For example, see S.N. Groudev et al. "Bioremediation of a Soil Contaminated with Radioactive Elements." *Hydrometallurgy*, Vol. 59#2-3 (February 2001), pp. 311–318; J.R. Lloyd and J.C. Remshaw. "Bioremediation of radioactive waste: radionuclide-microbe interactions in the laboratory and field-scale studies." *Current Opinion in Biotechnology*, Vol. 16 (May 2005), pp. 254–260. [online]. [cited December 20, 2012]. envismadrasuniv.org/Bioremediation/pdf/Bioremediation%20of%20radioactive%20waste.pdf.

7. Ingenium — College of Engineering Blog. "Radioactive Soil Remediation." Oregon State University, June 6, 2012. [online]. [cited December 20, 2012]. blogs.oregonstate.edu/engineering/2012/06/06/radioactive-soil-remediation/.

8. Ingham, Skype interview.

9. Josephine Porter Institute for Applied Bio-Dynamics, Inc. 2012. *Biodynamic Compound Preparation — Remedy for Radioactivity*. [online]. [cited March 13, 2013] jpibiodynamics.org/node/447.

10. Stamets, Skype interview.

## Chapter 11: Self-Care for Grassroots Remediators, Community Members and Disaster Responders

1. Guido Mase, phone interview, April 25, 2012.

2. Hiroko Furo. "Dietary Practice of Hiroshima/Nagasaki Atomic Bomb Survivors. "Wesleyan University, Illinois. [online]. [cited December 22, 2012]. yufoundation.org/furo.pdf; Meg Wolff. "Radiation and Miso's Hopeful Healing Powers." *The Huffington Post*, March 17, 2011. [online]. [cited December 22, 2012]. huffingtonpost.com/meg-wolff/radiation-misos-hopeful-h_b_836744.html; M. Ohara et al. "Radioprotective effects of miso (fermented soy bean paste) against radiation in B6C3F1 mice." *Hiroshima Journal of Medical Science*, Vol. 51#4 (December 2001), pp. 83–86. [online]. [cited December 22, 2012]. ncbi.nlm.nih.gov/pubmed/11833659; International Union of Food Science & Technology and

Institute of Food Technologists. *Radioactive Fallout from the 2011 Japan Nuclear Plant Accident and Some Recommended Precautions and Counter-measures.* [online]. [cited January 9, 2013]. worldfoodscience.org/cms/? pid=1006164.

3. W. Shurtleff and A. Aoyagi. *History of Miso, Soybean Jian (China), Jang (Korea) and Tauco/Taotjo (Indonesia) (200 BC–2009): Extensively Anno-tated Bibliography and Sourcebook.* SoyInfo Center, 2009, p. 941.

4. Seiichiro Yamamoto et al. "Soy, Isoflavones, and Breast Cancer Risk in Japan." Japan Public Health Center-Based Prospective Study on Cancer and Cardiovascular Diseases (JPHC Study) Group. *Journal of the National Cancer Institute,* Vol. 95#12 (June 18, 2003). [online]. [cited December 22, 2012]. nci.oxfordjournals.org/content/95/12/906.full.pdf.

5. Ryan Drum. *Sea Vegetables for Food and Medicine.* January 13, 2011. [on-line]. [cited December 22, 2012]. ryandrum.com/seaxpan1.html; St. Francis Herb Farm. *Bladderwrack.* [online]. [cited December 22, 2012]. stfrancisherbfarm.com/product.aspx?ID=113.

6. Bill Bodri. *How to Help Support the Body's Healing After Intense Radioactive or Radiation Exposure: The Medical, Naturopathic, Nutrional, Herbal, Commonsense External and Internal Approaches.* Top Shape Publishing, LLC., 2004. [online]. [cited December 22, 2012].alternativemedia.ca/ RadiationDetoxDraft.pdf.

7. S.C. Skoryna et al. "Studies on Inhibition of Intestinal Absorption of Radioactive Strontium." *Canadian Medical Association Journal,* Volume 191 (June 1964), pp. 285–288 cited in International Union of Food Science, *Radioactive Fallout from the 2011 Japan Nuclear Plant Accident.*

8. Bodri, *How to Help Support.*

9. US Department of Health and Human Services. "Dietary aspects of car-cinogenesis." November 1981 cited in Dr. Sat Dharam Kaur. "Nutritional Medicine for Protection from Radiation-Induced Damage." [online]. [cited December 22, 2012]. vitalitymagazine.com/article/nutritional-medicine/; Bodri, *How to Help Support.*

10. Bodri, *How to Help Support.*

11. L.P. Loseva and I.V. Dardynskaya. "Spirulina — natural sorbent of ra-dionucleides." 6th International Congress of Applied Algology, September 1993 quoted in Intergovernmental Institution for the Use of Micro-Algae Spirulina Against Malnutrition. *Publications and Reports.* [online]. [cited December 22, 2012]. iimsam.org/publications_and_reports.php.

12. Zhang Cheng-Wu, Tseng Chao-Tsi and Zhang Yuan Zhen. "The effect of polysaccharide and phycocyanin from *Spirulina platensis* var. on peripheral blood and hematopoietic system of bone marrow in mice." Paper presented at the 2nd Asia-Pacific Conference on Algal Biotechnology, Malaysia, 1994.

13. Matsubara et al. "Radioprotective effect of metallo-thionine." Presented at Radial Rays conference, Tokyo, Japan, 1985; Amha Belay et al. "Current Knowledge on Potential Health Benefits of *Spirulina*." *Journal of Applied Phycology*, Vol. 5#2 (April 1993), pp. 235–241. [online]. [cited December 22, 2012]. link.springer.com/article/10.1007%2FBF00004024.

14. Bodri, *How to Help Support*.

15. Steven R. Schechter. *Fighting Radiation and Chemical Pollutants with Foods, Herbs, & Vitamins: Documented Natural Remedies That Boost Your Immunity and Detoxify*, 2nd ed. Vitality, 1992.

16. Bodri, *How to Help Support*.

## Conclusion: Final Words for a Fertile Way Forward

1. Wendell Berry. "Work Song, Part 2: A Vision" in *New Collected Poems*. Counterpoint, 2012, pp. 217–218.

## Appendix 1: Contaminants 101

1. For more information on contaminants, search the following websites:

   Cleanuplevels.com. *Cleanup/Screening Levels For Hazardous Waste Sites*. [online]. [cited December 14, 2012]. cleanuplevels.com.

   Eco-USA.net. *Chemicals*. [online]. [cited December 14, 2012]. eco-usa.net/toxics/chemicals/index.shtml.

   Environment Canada. *National Pollutant Release Inventory*. [online]. [cited December 14, 2012]. ec.gc.ca/inrp-npri/.

   US Agency for Toxic Substances and Disease Registry. *Minimal Risk Levels (MRLs) for Hazardous Substances*. [online]. [cited December 14, 2012]. atsdr.cdc.gov/mrls/mrllist.asp#17tag.

   US Environmental Protection Agency. *Pesticides: Health and Safety*. [online]. [cited Decenmber 14, 2012]. epa.gov/pesticides/health/human.htm.

   US Environmental Protection Agency. Contaminated Site Clean-Up Information. [online]. [cited December 14, 2012]. clu-in.org/contaminant focus/.

2. J. Conant and P. Fadem. *Community Guide to Environmental Health*. Hesperian Foundation, 2008, p. 341. [online]. [cited December 21, 2012]. en.hesperian.org/hhg/A_Community_Guide_to_Environmental_Health.

3. US Geological Survey. "Volatile Organic Compounds: Definitions." USGS Toxic Substances Hydrology Program, 2005. [online]. [cited December 21, 2012]. toxics.usgs.gov/definitions/vocs.html .

## Appendix 2: Conventional Remediation Techniques

1. For further information, consult US Environmental Protection Agency, Contaminated Site Clean-Up Information. *Technologies and Remediation.* [online]. [cited January 9, 2013]. clu-in.org/remediation/. The US EPA Citizen Guides for different conventional clean-up technologies can be downloaded from cluin.org/products/citguide/. Also check the US Federal Remediation Roundtable website: frtr.gov/default.htm.

# Index

# About the Author

L EILA DARWISH is a community
organizer, urban gardener and permaculture rabble rouser with a degree in
Environmental Conservation Sciences.
Most of her grassroots organizing has
centered on environmental justice issues
in communities struggling with either
the threat of or the enduring legacy of
toxic contamination of their land and
drinking water. Her focus on grassroots
bioremediation stems from a deep commitment to justice and the passionate
desire to empower people by providing
them with simple, practical, accessible
and transformative tools for regenerative earth repair.

CREDIT: SARA DENT

If you have enjoyed *Earth Repair*, you might also enjoy other

# BOOKS TO BUILD A NEW SOCIETY

Our books provide positive solutions for people who want to make a difference. We specialize in:

**Sustainable Living • Green Building • Peak Oil
Renewable Energy • Environment & Economy
Natural Building & Appropriate Technology
Progressive Leadership • Resistance and Community
Educational & Parenting Resources**

For a full list of NSP's titles, please call 1-800-567-6772 or check out our website at:

**www.newsociety.com**

new society
PUBLISHERS